Physicists on Wall Street and Other Essays
on Science and Society

Jeremy Bernstein

Physicists on Wall Street and Other Essays on Science and Society

 Springer

Jeremy Bernstein
New York
NY, USA

ISBN: 978-0-387-76505-1   e-ISBN: 978-0-387-76506-8
DOI: 10.1007/978-0-387-76506-8

Library of Congress Control Number: 2008931402

Printed on acid-free paper

9 8 7 6 5 4 3 2 1

springer.com

About the Author

Jeremy Bernstein has had a long and distinguished career in which he made major contributions in the fields of writing, teaching, and science. He is currently a professor emeritus of physics at the Stevens Institute of Technology in Hoboken, New Jersey. He was a staff writer for the *New Yorker* magazine from 1961 to 1995 and has written more than a dozen books on popular science and travel. He has won the AAS-Westinghouse Prize and (twice) the U.S. Steel-American Institute of Physics science writing prize. His book *Albert Einstein* was nominated for a National Book Award. He is also the recipient of the Brandeis Creative Arts Medal for non-fiction, the Britannica Award for the dissemination of knowledge, and the Germant Award, given by the American Institute of Physics for contributions that link physics to the arts and humanities. He has taught non-fiction writing at Princeton University. Professor Bernstein was born in Rochester, New York and was educated at Harvard University, where he received three degrees. His primary interests in physics research are in the areas of elementary particles and cosmology.

Preface

Everyone has their own way of learning. Mine has always been verbal. When I was a kid, I used to think out loud. I was caught at it once by a school janitor—a very nice man who suggested that what I was doing was a little strange. After that, whenever I wanted to learn something new or complicated, I'd begin by reading up on it and then lecturing to myself in my head. I still do this. If I get stuck I realize that I had better do some more reading or maybe ask an expert. Once I think I have it down pretty well, I write an essay. Almost never are these essays commissioned, and almost never do I know where, if anywhere, they are going to be published. For this reason a lot of my essays are too long to be published in most magazines or newspapers. It is very rare that you will get as much a 3,000 words in a magazine. The old *New Yorker*, for which I wrote for thirty-five years, was different. Many of the things I turned in emerged substantially longer than what they were in their original form. But this is incredibly rare. If the editing is good, an editor can preserve the essence of what you are saying while chopping the piece in half, or even more than half. Like most authors, I have learned to accept this if I respect the editor and the publication. But there is always a sense of loss.

There is a bit more opportunity with books of essays to take the time to get where you are going with an idea. The essays in this book are about science and scientists in a very broad sense. Each one reflects something about which I became extremely interested at various times. A couple of these essays, the ones that were published in trade journals, were published at their original length or longer. The rest have either never been published at all or have only been published partially. The subject matter is very eclectic. I hope the reader will find the choice interesting.

Jeremy Bernstein

Acknowledgments

"Options" first appeared in a somewhat different form in *Commentary* and "Heisenberg in Poland" and "Orion" did likewise in *The American Journal of Physics*.

Contents

About the Author v

Preface.. vii

Acknowledgments ix

Part I   Economists

1   Options 3

2   Black-Scholes.. 9

3   The Rise and Fall of the Quants 15

Part II   Scientists

4   Heisenberg in Poland.. 25

5   The Orion.. 35

6   Tales from South Africa... 43

7   A Nuclear Supermarket 47

8   Ottavio Baldi: The Life and Times of Sir Henry Wotton 57

Part III   Linguists

9   The Spencers of Althorp and Sir William Jones: A Love Story 73

10   All That Glitters 87

11   In a Word, "Lions".. 107

Part IV   Fiction and Stranger than Fiction

12   The Pianist, Fiction and Non-fiction... 113

13   Rocket Science... 121

14   The Science of Michel Thomas 133

15   Topology.. 147

16   What the #$*!? 155

Author's Note About "Beating the System" 159

Index... 169

Part I
Economists

Chapter 1
Options

Da könnt' mir halt der liebe Gott leid tun, die Theorie stimmt doch-Then I would have been sorry for the dear Lord—the theory is correct. (Einstein's response to a student who asked him what his feelings would be if experiment failed to confirm his theory of gravitation)
    If you decide you don't have to get A's, you can learn an enormous amount in college.

—I. I. Rabi

In the spring of 1969 I got the somewhat lunatic idea that I wanted to go to the northwest frontier of Pakistan to see the high mountains—K-2, Nanga Parbat, and the like. As it happened, I had a colleague in physics who was a Pakistani, and he had a connection both with the University of Islamabad and the Ford Foundation, which had a program to send scholars to Pakistan to teach at the university. He arranged for me to become a Ford Foundation visiting professor. I told the foundation that I intended to drive to Pakistan and, after expressing considerable surprise, they agreed to give me what they would have had to spend on first-class airfare towards purchasing a specially modified Land Rover suitable for making such a trip. I then persuaded two friends to come with me and in something less than a month we drove to Pakistan by way of Greece, Turkey, Iran, and Afghanistan and over the Khyber Pass to Islamabad. Upon arriving, I learned that the university was closed for a month as a kind of punishment for some sort of student uprising. This gave us a full month to explore the frontier, places such as Chitral, Swat, Hunza, and Gilgit—now inaccessible, as is the rest of our route. At the end of the month my friends went back to France, and I moved into the Ford Foundation staff house in Islamabad to take up my teaching duties.

It was a pleasant, if somewhat lonely, existence as the sole resident of the staff house, apart from the staff, which included a driver for the car I had been assigned. But after about a month, one morning I heard a pair of English-speaking voices, male and female. Upon investigation it turned out to be another Ford professor and his wife. But this was not *any* professor. It was Marshall Stone. Although I had never met him, Stone was one of my heroes. He was one of the world's greatest mathematicians. He had taught at Harvard for many years and then, in 1946, he was brought to the University of Chicago to create what became the leading school of mathematics in this country. Moreover, Stone was the teacher of my teacher at Harvard, George Mackey, who had interested me in the mathematical foundations of the quantum theory, some of which had been provided by Stone. Stone, who died in India in 1989 at the age of 85, was, incidentally, the son of the late Chief Justice of the Supreme Court, Harlan Fiske Stone. Now, here he was, in the Ford Foundation staff house accompanied by his rather recently acquired wife, Vila, a very attractive and voluble Yugoslavian.

It turned out that Stone, who was an inveterate traveler, had also come to Pakistan to visit the frontier. Thus, the three of us spend a good deal of time together, during which I told him what I knew about it. In the course of this, Vila mentioned that she had a daughter who lived in New York and that I might like to meet her. It took some time, but eventually I called. She knew who I was, and we became friends. Some years later, she told me that she was seeing someone with whom I might have things to talk about, since he was studying "derivatives." A derivative in calculus is the rate of change along a curve. It is one of the first things one learns in calculus. I assumed that this fellow was taking a first course in calculus, which did not seem to me to be much of a conversation piece. In actuality, he turned out to be an amiable chap whose name was Myron. I forget what we talked about, but I managed to steer the conversation away from any mention of derivatives. I have a dim recollection that, at one point, his date whispered in my ear that Myron was going to win the Nobel Prize someday. I am sure that I thought that if they were giving out Nobel Prizes for learning calculus, things were worse than I had imagined. But, as it turned out, she was right. In 1997, not long after our meeting, Myron Scholes and Robert Merton shared the Nobel Prize in economics. Scholes and Fischer Black, who had died two years earlier, had created what is known as the Black-Scholes equation, which they published in 1973. (Merton invented another approach to the same problem. We will attempt to explain all of this later.) It does indeed deal with derivatives—investment vehicles such as options on stocks or bonds, whose present value is derived from the projected future values of the financial commodities—stocks and bonds—that underlie them. The Black-Scholes equation, and its many adumbrations, is used to set the price of such options. It is, If you like, the Newton's Law or the Schrödinger equation of the whole field of financial engineering that makes these derivative markets operate.

I had more or less forgotten about all of this until I read an autobiographical memoir entitled *My Life as a Quant; Reflections on Physics and Finance* by Emanuel Derman. A "quant" is the rubric used on Wall Street and elsewhere to describe people who practice quantitative financial analysis—financial engineering—for which the Black-Scholes equation is a prototype. Physics comes into Derman's memoir because, although he became a professor in Columbia University's Department of Industrial Engineering and Operations Research and ran the financial engineering program there, he actually had a Ph.D. in physics from Columbia, which he obtained in 1973. He was one of the early POWS—Physicists on Wall Street—having joined the financial firm of Goldman Sachs in 1985. The first part of his book traces the somewhat unlikely steps that took him from his native Cape

Town, South Africa, first to Columbia and then via the AT&T Bell Labs, and elsewhere, to Wall Street. For reasons that will be explained later, the stops along his route are especially familiar to me. Indeed, although I do not have any specific memories, Delman notes in his book that our paths crossed at various times. He and I are both theoretical elementary particle physicists, and our world is not large.

Derman arrived in New York in 1966. He had taken an undergraduate degree in applied mathematics and theoretical physics at the University of Cape Town. As he notes, it is quite unusual to find a university in which someone could do an undergraduate degree in physics and have, apart from a one-year lab course, very little contact with experiments. Upon graduation he applied for scholarships to study in various universities outside South Africa and obtained one from Columbia. The physics department he found at Columbia in the 1960s was very familiar to me. I spent a lot of time there and at one point even had a visiting appointment. In Derman's day the department was still under the aegis of the noted Nobelist I. I. Rabi, whose standards were extremely high and, although Derman did not see this side of his character, could be very tough. To give an example; the economist Milton Friedman's son David was a post doc in the department. I have an ineluctable memory of a departmental Chinese lunch during which young Friedman was heard discoursing on every subject known to man. Rabi suddenly said, "Be quiet. We'll hear from you when you are older." That was the end of that. Rabi wasn't the only present and future Nobelist in that department. You had to be very good, and very determined, to survive in that atmosphere.

Because of his course background, Derman was at a disadvantage. Although he passed the qualifying examinations with high enough scores so that he had the option of working in theoretical physics—which, for some reason, was reserved only to students who were considered especially bright—it was decided that he did not have enough knowledge of modern physics (things like quantum mechanics and its applications) to begin a thesis. He had to spend two years taking courses in subjects that most of the other graduate students already knew. Once having completed this he could begin to look for a thesis advisor. The leading theorist in the department was Tsung Dao Lee, known universally as "T. D." T. D. was in his early forties when Derman arrived. He had shared the Nobel Prize with his collaborator, Chen-Ning Yang, a decade earlier. Derman had a shrewd eye for people. Here is what he wrote about T. D.

"With his Moses-by-Michelangelo persona, beams of light emerging from his forehead, T. D. radiated an intense purity. At first I imagined that his rigorous questioning was the by-product of a pure search for knowledge and truth. Later I thought I began to detect a latent glee with which he savaged the imperfections in other people's talks. He enjoyed disorienting them."[1] I am not so sure that it was a matter of enjoyment. T. D. simply could not stand listening to something he was sure was wrong. On the matter of a "pure search for knowledge" I once put the following proposition to a small group of physicists at lunch. I said suppose God

[1] We will refer to Derman's book *My Life as a Quant*, Wiley, Hoboken, 2004, hereafter as M. L. This quote can be found on p. 35 of M. L.

offered you a book of his own composition in which all the problems of physics were solved. Would you look at it knowing that it would end your role in discovery? Many of the physicists said they would. T. D. was not sure.

Few theoretical physics students at Columbia worked with T. D because T. D. accepted very few students—only the *wunderkinder*—so Derman did not try. He describes how he tried to connect with one of T. D.'s ex *wunderkind*, Gerald Feinberg. Some of the *wunderkinder* became faculty members at Columbia, and Feinberg was one. Derman's unsuccessful attempts to approach Feinberg are amusing to read, especially to me, as he was my best friend until his death in 1992 at the age of 58. I don't know how Derman's life would have turned out if he had been able to work with Feinberg, but he became the first graduate student of another of the *wunderkinder*, Norman Christ. Derman writes that because he and Christ were about the same age, surprisingly perhaps, he never found a comfortable way to relate to him; he was never even on a first-name basis with him. Nonetheless, he was able to produce a thesis, although this took another five years. He notes wryly that about 10 percent of his projected life span was spent getting a Ph.D. degree at Columbia. He wrote his thesis on what we refer to as "phenomenology"—applying some underlying theory to make a model that either predicts or explains an experimental result. In this instance one of the creators of the underlying theory was Steven Weinberg, who shared a Nobel Prize for this work. From the sound of it, Derman wrote a very respectable thesis that required him to learn to use the rather primitive computer facilities that were then available. One had to program machines with IBM punch cards—a very tedious and error-prone exercise. The thesis was good enough to get him a post-doctoral position at the University of Pennsylvania. In the meanwhile, he had married a young woman who had left Czechoslovakia after the Russian occupation. She was then also studying physics but later switched to biology.

The next several years were very difficult for the Dermans. Emanuel moved from one temporary job to another, usually in cities far from where his wife was. One of these jobs was in New York at the Rockefeller University. Life at the Rockefeller was pretty lush, and this time he was in the same city as his wife. However it became clear after his second year that his appointment was not going to be renewed. He even thought of giving up physics and going to medical school, but he could not quite bring himself to do it. Finally, he took an assistant professorship at the University of Colorado in Boulder. This was a real nightmare, since he was now separated geographically from both his wife and very young son. After one year he had had enough and took a job in the Business Analysis Systems Center at Bell Labs in Murray Hill, New Jersey, to which he could commute from New York. Derman's description of Bell Labs as he experienced it was surprising to me. He hated the place. His chapter on Bell Labs is called "In the Penal Colony," a reference to the Kafkaesque petty bureaucracy and enforced regimentation that he found. The reason that I was surprised was that in the period that we are discussing—the early 1980s—I spent a good deal of time at the labs, although I did not meet Derman there.

Indeed, I was not going there to do physics but rather to write about the place. I wrote a series of linked profiles of people at the laboratory. In 1984, it was published as a small book, *Three Degrees Above Zero*.[2] This is a reference to the temperature of the background radiation left over from the Big Bang, which was discovered in 1964 serendipitously, by two Bell Labs physicists, Arno Penzias and Robert Wilson, who received the Nobel Prize in 1978 for this work. I interviewed them as well as the third Nobelist at the labs, Philip Anderson. These three were part of a long tradition of Bell Labs scientists who had won the Nobel Prize, including William Shockley, Walter Brattain, and John Bardeen, all of whom shared the 1956 Nobel Prize for their discovery of the transistor. The people that I interviewed were at the top of their fields—people whom any university would have been delighted to have. At the labs they could do pretty much what they wanted. For them, Bell Labs was the real ivory tower, a first-rate research facility where in those pre-divestiture days—AT&T was broken up in January of 1984, and some of the Labs eventually became Lucent and the rest went with the Baby Bells—there was no concern about scrounging for money for research projects. Of all the people that I interviewed there was only one overlap with the people that Derman knew. This was Ken Thompson, a computer genius who, along with another Bell Labs computer genius, Dennis Ritchie, created the UNIX operating system. This is the multi-user, multi-task system that is used to run the computer complexes in most centers around the world. To run the system they created a language called "C," which, with its variants, became the language of choice for programmers. Thompson, along with a Bell Labs physicist named Joe Condon, had, at the time I met them, constructed a dedicated chess-playing machine they called "Belle." This was not a program but a machine that was hard-wired to play chess. It played just below the Grand Master level and was, at the time, the champion chess machine. As part of my interview, I was given the chance to play it—losing gracefully. Derman would have liked to join a research group with people like Thompson and Ritchie, but all his requests were refused. By the early 1980s he had had enough.

As it happened, this coincided with the time in which the major brokerage firms on Wall Street and elsewhere were building up their financial engineering departments. They were headhunting in places such as Bell Labs in search of potential quants. This had to do with a change in the brokerage business from merely selling stocks and bonds to dealing in all sorts of derivatives. For example, Salomon had put together a very powerful group of such analysts under the aegis of one John Meriwether (More about him later). One of the consultants to this group was Robert Merton, the Harvard professor who later shared the Nobel Prize with Scholes. Incidentally, it was Merton's father (also Robert but with a different middle initial), a noted sociologist of science at Columbia University, who coined the phrase "self-fulfilling prophecy," something that might well have characterized the later activities of his son and his collaborators. Even the staid brokerage firm of Goldman, Sachs was adding quants. In December of 1985, Derman took a job at Goldman in the Financial Strategies Group, where he had his first encounter with Black-Scholes.

---

[2] *Three Degrees Above Zero* by Jeremy Bernstein, Mentor Books, New American Library, New York, 1986.

Chapter 2
Black-Scholes

> Through my parents and relatives I became interested in
> economics and, in particular, finance. My mother loved busi-
> ness and wanted me to work with her brother in his book pub-
> lishing and promotion business. During my teenage years, I was
> always treasurer of my various clubs; I traded extensively
> among my friends; I gambled to understand probabilities and
> risks; and worked with my uncles to understand their business
> activities. I invested in the stock market while in high school
> and university. I was fascinated with the determinants of the level
> of stock prices. I spent long hours reading reports and books to
> glean the secrets of successful investing, but, alas, to no avail.

—Myron Scholes, Nobel autobiography, 1997

At this point let us interrupt the narrative in order to explain the Black-Scholes-
Merton revolution. Otherwise you will not be able to make much sense out of what
follows. Derman gives a good qualitative discussion of this, but, as the great
nineteenth-century Scottish physicist James Clerk Maxwell used to say, "I didn't
see the 'go' of it. " If you put "Black-Scholes" into Google you will find something
like 1.46 million entries. Most of them are technical, proposing solutions to the
equations or trying to generalize or derive them. Some of these sites have clearly
been posted by ex-physicists, who note, for example, that the Black-Scholes equation can be morphed into the equation that describes the flow of heat. (We will
explain this, too.) There are offers to tutor you for a considerable fee. While wan-
dering through this jungle I came across the perfect site for my purposes. It is called
"Black-Scholes the Easy Way. " You can find it at http://homepage.mac.com/j.norstad.
The person who put it up, John Norstad, is a computer scientist who was learning
this stuff as a hobby. His posted notes represent his own learning process and are
very clear. We will use his examples in case someone wants to consult the site for
more details. But what is the basic problem?

At this time, as mentioned, financial institutions were doing a substantial
business in the sale of derivatives. A typical example is a stock option. This is a
contract between two parties that allows the buyer of the option to purchase a particular stock at a future time from the seller of the option at a contractually specified
price called the "strike price, " which is often but not always the price of the stock

J. Bernstein, *Physicists on Wall Street and Other Essays on Science and Society*,
© Springer Science+Business Media, LLC 2008

when you buy the option. Until that future time you do not own the stock. You own an option *to buy* the stock at the fixed price. If the stock has gone up by the time you buy it in the future you are, using the term of choice, "in the money. " If the stock goes down, you don't buy it but are out the cost of the option— "out of the money. " The question is, what should the price of the option be when you buy it? This is what the Black-Scholes equation purports to allow you to compute.

To see what is involved we will, following Norstad, consider a "toy" model. This illustrates many of the general features of the problem without the mathematical complexity.

In the toy model there is a stock whose current value is $100, which, in this example, will be the strike price. What makes the model a toy is that at the time the option is to be exercised there are only two possible prices; $120 and $80. In the real world there will be a continuum of prices, which is what Black-Scholes must deal with. The kind of option we are considering here is called a "European call option. " It can only be exercised at one definite time in the future. An "American call option" can be exercised at any time. We will further assume that the probability of the stock's rising to $120 is three quarters, while the probability of its falling to $80 is one quarter. What then should you be willing to pay for the option? At first sight this seems like a simple question to answer. With these probabilities the expected outcome is $\frac{3}{4} \times \$20 + \frac{1}{4} \times \$0$, which is $15. Thus one might assume that the option should be worth $15 to you and that you can then, with a high probability, expect to earn $5 if you buy the option. This would be true if we were not able to engage in financial engineering—an activity that goes under the rubric of "arbitrage. " With arbitrage one can gain a certain profit with no risk at all. The cost of this arbitrage is what determines the cost of the option. This changes everything and explains why the financial institutions were hiring quants by the carload. Here is how the arbitrage works in this case.

Let's assume there is a friendly bank that is willing to lend money interest free. This is a simplification in the analysis that can readily be corrected. If you want to see how including interest modifies the results in the toy model, you can look at Norstad's website. Only a little high school algebra is required. We do not lose any matters of principle if we make this assumption. Likewise we assume that we can buy fractional shares of this stock from a friendly broker commission-free. To use another physics/economics term, we make the problem "frictionless. " In physics a friction force generates heat without doing any useful work. Here, the friction generates money loss without helping us with the bottom line.

Now suppose you have made the expectation calculation given above and are willing to give $15 to buy the option. We will now see how we can, using arbitrage, always pocket $5 no matter what the outcome is. To achieve this we take the $15 from you and put $5 in our pocket. You will never see the fiver again. We then borrow $40 from the friendly bank-"leverage. " We take your $10 and the borrowed $40 and buy a half share of the stock. This is called the "hedge. " In this case it has cost us $10 to replicate the option. This will turn out to be its true value.

Now we have the two cases. If the final price is a $120, you will exercise your option to buy the stock at a $100. We are obliged to deliver the stock to you at that price. What we will do is to sell our half-share for $60, repay the bank its $40, and add the remaining $20 to the $100 you gave us to buy the share so we can give it to you. We have, of course, pocketed the $5. If, on the other hand, the final price is $80 you will not exercise your option. If you really like the stock you can simply buy it outright for $80. We will sell our half-share for $40, which we will then return to the bank, still pocketing the $5. This means that if you have given us $15 for the option you have overpaid by $5. If you think about it, you will see that the $10 price is a kind of tipping point. If we can sell the option for more than $10 we will make money, and if someone wants to sell us the option for less than $10 we will buy it and again make money. The $10 is a kind of equilibrium price at which it is not profitable to either buy or sell. Black and Scholes approached the problem of option evaluation using arbitrage as an equilibrium problem—you need to find the price at which there is equilibrium between buying and selling. Merton had a different approach, and this is the one that is now more generally used.

To understand it, note that in finding the correct option price in the presence of the possibility of arbitrage, the probabilities of three-quarters and one quarter played no role. These probabilities only entered when, in the absence of arbitrage, we used them to compute the expected gain. We never had to use them when we found the cost of the hedge—which is the correct option price. This is important because in real life there is little likelihood that we would ever be given these probabilities in any reliable way. In fact, in an important sense, the presence of a buyer of the option is irrelevant. Suppose we just construct a portfolio that consists of $10, and a $40 loan from the bank, which we then invest in half a share of the stock which is selling for $100 a share. This is called a "synthetic option. " You can easily persuade yourself that if we sell this stock when its value is either $120 or $80, the amount that we gain or lose is the same as what the buyer of the option gains or loses in the preceding example if he or she pays $10 for the option. The essence of Merton's approach is to show that one can, in general, construct synthetic options that have the same outcome as the real options. The cost of the synthetic option is the same as the cost of the real option and, by what is called the Law of One Price, this *is* the cost of the option. This is what these brokerage firms do—they construct synthetic options.

In discussing the equivalence of these two methods of pricing options, Derman was reminded of what happened in quantum electrodynamics in the late 1940s. There were two approaches. Julian Schwinger started from first principles and by carrying out a series of horrendous calculations—which, as Oppenheimer said, only Schwinger could have done—produced numbers for physical quantities that could be compared to experiment. Richard Feynman, on the other hand, arrived at these numbers by using pictorial methods and intuitive arguments. A reasonably competent graduate student could learn them in a few days. I remember the disquiet I felt when I first studied Feynman's papers. Why did these tricks work? Indeed, did they always work? The matter was laid to rest when Freeman Dyson in a mathematical tour de force showed that Schwinger and Feynman had found two equivalent representations of the same theory. You could use Feynman's Mozartean calculus knowing that Schwinger's Bach-like logic made it legitimate.

What Black, Scholes, and Merton had to confront was the fact that in the real world we do not have a situation in which there are just two future prices but in fact a continuum. This gets one into the question of how you can predict the future of a stock price. To deal with this Black and Scholes adopted a model that supposed that stock prices follow what is known as a "random walk." We will explain this, but first remarkably the same model was used by a French mathematician named Louis Bachelier to derive the price of what is known as a "barrier option," an option that is extinguished if the stock price crosses a certain barrier. (One feature of Bachelier's model that was not realistic was his use of negative as well as positive stock prices.)[3] This work was contained in his Ph.D. thesis, *Théorie de la Spéculation*, which he presented at the Sorbonne in 1900! Although Bachelier wrote both books and papers on this kind of probability theory, his work was not much appreciated, even in France. A great deal of it was rediscovered and made more rigorous by people such as Norbert Weiner. It is now embedded in what is known as the "stochastic calculus."

In 1905, Einstein, who had not heard of Bachelier, used these ideas to analyze what is known as "Brownian motion." In 1827, the Scottish botanist Robert Brown made a very odd discovery. He noticed that if pollen grains, which could only be seen though a microscope, were suspended in water they executed a curious dancing motion. Brown made the natural assumption that these pollens were alive, but then he tried other microscopic particles made of things that were clearly not alive and found the same effect. By the end of the century the correct explanation had been guessed at; namely, that these very small particles were being bombarded by the still smaller—indeed invisible—water molecules and that the "dance" was in response to these collisions. In 1905, Einstein showed that this assumption had precise and measurable consequences. The same work was done independently at about the same time by the Polish physicist Marian Smoluchowski. The basic idea can be illustrated by what is known as the "drunkard's walk."

A drunkard begins his walk at, say, a lamp post. At each step he can go, say, 2 feet, but in a totally random direction. The question then is how far on the average the drunkard will go, as the crow flies, away from the lamp post after a given number of these steps, say N. When first confronted with this question many people are tempted to say the drunk won't get anywhere, since he will simply retrace each step. But a moment's reflection convinces one that this is essentially impossible. After each step the drunkard's new direction is totally random.[4] The path will appear jagged, but the distance from the lamp post continues to increase. Indeed, the fundamental result of this analysis is that the average distance increases as the square root of the number of steps N.[5] This square root feature shows up in the

---

[3] Bachelier's random-walk model predicted that the spread in stock prices would increase as the square root of the time. But there was no limit in his model as to how low a stock could go. It could become negative, which means the company would be paying you to buy the stock. This flaw in Bachelier was first noted by the economist Paul Samuelson.

[4] A special case is a one-dimensional random walk. After each step the drunkard is equally likely to go forward or backward. Nonetheless he will move inexorably away from the lamp post. This example makes it clear why it takes longer to reach a certain average distance than it would if one just walked there.

[5] To be precise we are talking about the root mean square distance. This is the square root of the average of the square of the distance.

Brownian motion. Here the "drunkard" is a grain pollen that is being driven hither and yon by its collisions with the invisible water molecules. The distance that the pollen goes is, as Einstein showed, proportional to the square root of the time during which you observe it. In his papers Einstein put in a few numbers for various substances at room temperature. He predicted that typically such a suspended object might go some thousandths of a centimeter in a second. Experiments were done and, from the results, one could deduce the constants that went into his formula—constants that included the number of molecules in a cubic centimeter. These experiments convinced most skeptics that atoms really existed. They did not convince the physicist-philosopher Ernst Mach, who liked asking atomists, "Have you seen one? "

Assuming that stock prices follow a continuous random walk, Black and Scholes could make a prediction for the future distribution of the price of a stock. In Derman's book he has a typical graph that plots this. Prices on the graph deviate from the initial price, making a kind of wedge with the pointed end at the initial price. The wedge continually widens as time goes on, so that the price becomes more uncertain. Knowing the probable future prices of the stock, Black and Scholes were able to derive an equation for the value of the option at any given time. It is a differential equation—one involving the sort of derivatives that I mistakenly thought that Scholes was first learning about when I met him. Since the value of the stock is constantly changing—unlike our toy example, where it changed only once—the hedge must also be constantly adjusted. In the general case, the price of the option is the total price of this constantly adjusted hedge. It is this hedge price that is used by the people who sell these options.

The transport of heat is also such a diffusion process in which the faster moving "hot" molecules transfer their momentum to the cooler particles until the two reach an equilibrium. Therefore it is not entirely surprising that one can mathematically transform the Black-Scholes equation into the equation that describes heat diffusion. This equation has been studied for well over a century so that there are a lot of mathematical tools available. Indeed, in their paper, Black and Scholes transform their equation into the heat equation, which they then solve.

So why would I, as a physicist, find the Black-Scholes model quite odd? All physical theories are models. For example, quantum electrodynamics, which is the most precise theory ever created, operates in a model universe that contains only electrons and quanta of light-photons. The rest of the universe, with its neutrons, protons and mesons, quarks, and the like, are ignored. But the object of this model, like all the other models in physics, is to predict the future. If the model is correct, then the numbers and curves one calculates with it should be confirmed by experiment. But the Black-Scholes model is quite different. It uses a model of the future to describe the *present*.[6] In the absence of this, or some equivalent model, present

---

[6] This is reflected in the conditions under which the equation is solved. What is specified is the *future* value of the option at the time it is exercised. This is either zero or the strike price less the amount you have borrowed depending on how the stock's price has evolved. When you solve a physics equation you generally use data from the present to find your solution, which then allows you to predict the future.

stock options have no reasonable assigned value. Presumably the test of the model is that if one uses it as a guide to buy these options and, as a result, goes broke, one would be inclined to re-examine the assumptions. But we have digressed. Let us continue the discussion of quants in the next essay.

Chapter 3
The Rise and Fall of the Quants

Markets can remain irrational longer than you can remain solvent.

—John Maynard Keynes

By the time Derman came to Goldman, Sachs in 1985, the use of the Black-Scholes equation to evaluate options had become commonplace. However, the equation had gotten off to a somewhat rocky start. In 1968, Scholes became an Assistant Professor of Finance at MIT in the Sloan School of Management. Black was a consultant for the Arthur D. Little Company in Cambridge. The two of them began collaborating on various economics problems. But at MIT there was also Paul Samuelson, one of the creators of modern mathematical economics. Samuelson had studied the use of Brownian motion to predict stock prices and two of his students, Richard Kruizinga and A. James Boness, had written theses attempting to use the model to derive values for options. Indeed, Kruizenga had cited Bachelier's work, so it was familiar at least at MIT. But it was left to Black and Scholes to finish the job. They first derived their equation in 1969. They submitted their results in the fall of 1970 to the *Journal of Political Economy*, which rejected their paper. Next they tried the *Review of Economics and Statistics*, which also rejected their paper. They revised and simplified it. They then sent it back to the *Journal of Political Economy*, which finally published it in the May/June 1973 edition under the title "Pricing of Options and Corporate Liabilities."[7] Merton published his somewhat more general paper, "Theory of Rational Option Pricing," in which he referred to Black and Scholes as well as Bachelier, in the *Bell Journal of Economics and Management Science* at about the same time.[8] They turned out to be two of the most influential papers ever published in economic theory. In light of that it is amusing to read at the beginning of Merton's paper his disclaimer that "Because options are specialized and relatively unimportant financial securities, the amount of time and space devoted to the development of a pricing theory might be questioned."[9]

---

[7] Vol. 81, 3, pp. 637–654.

[8] 4/Spring, pp. 141–182.

[9] Merton, op. cit., p. 141.

J. Bernstein, *Physicists on Wall Street and Other Essays on Science and Society,*
© Springer Science+Business Media, LLC 2008

The original work was done on stock options. By the 1980s the problem had become how to extend it to bond options. At any given time a bond will have a certain value. However, at the time of its expiration date, its value returns to the value at which it was issued—par. However, the expiration date could be many years off, 30, for example. Thus an options buyer might want to buy an option to buy the bond at some future date before the expiration. At the time when the option holder wants to exercise the option, the bond's price may differ from the price that was obtained when the option was purchased. Just as in the stock case, if the bond gains in value the owner of the option is in the money. Otherwise the option will not be exercised and the option buyer is out the cost of the option. In their first attempt at doing bond option valuation Goldman tried to use the Black-Scholes equation. By the time Derman came to the firm, it was already understood that the equation had a limited validity.

There were two important reasons. In the first place, future bond prices follow a different curve. This has to do with the fact that at the expiration date the bond price returns to its initial offering price. Thus the spread of future bond prices has a more banana-like shape than a wedge shape. It only resembles, at least vaguely, the wedge for short times. The second important difference is that bond prices tend to be connected to each other. Familiar examples are the Treasury bonds of different durations. Their prices tend to move in concert when interest rates change. This does not happen with stocks whose prices vary independently. The first problem could be dealt with by focusing on the yield, the average annual percentage return of the bond purchased at its present price and then held to maturity. The closer to maturity that you buy the bond the less relevant is the yield. There isn't enough time left to generate any appreciable yield before the bond matures. Therefore, even though Black-Scholes does not model this part of the curve very well, it does not matter very much. The problem of the interconnection of the bonds was much more serious. Fischer Black, who joined Goldman in 1984, the year before Derman arrived, invited Derman to collaborate with him and another quant named Bill Toy to attempt to make a new model for pricing bond options, one which would take these interconnections into account.

Fischer Black is certainly the hero of Derman's book. It was typical that, after he was made a partner at the firm in 1986, Black liked to point with pride to the fact that, of all the partners, he had the fewest shares in the firm. Having money was never one of his major interests. He was one of the first academics to be hired on Wall Street. Then Chairman Robert Rubin, who was interested in options trading, brought him in to Goldman's. Black had been teaching first at the University of Chicago and then at MIT. He had graduated from Harvard in 1959 as a physics major, but then switched to applied mathematics, taking his Ph.D. in 1964. His interest in economics came about when he began consulting for Arthur B. Little. His great strength was his lucidity. He did not like clutter, mental or otherwise. He refused to use a computer mouse because he felt that "mice" and other such pointing devices were an extraneous source of clutter. This meant that he had to introduce new definitions on his keyboard so he could navigate the screen. He also did not like graphs as opposed to numerical tables. These tables also required special prep-aration, since he did not countenance "trailing zeros, " numbers like 6.0000 where the zeros meant nothing or, still worse, had a misleading implication about precision. Derman found that if it had a specific question, Black was very ready to try to answer it, but otherwise, he was not very responsive. Derman writes,

"Fischer was precise and organized, quite punctilious. Every day he ordered the same ascetically healthy meal delivered to his desk. He liked to wear a Casio information-storing watch, which prompted some of his employee-admirers to do likewise. In his office giving audience, if you said something he found useful he'd write it down with his fine-pointed mechanical pencil on a fresh sheet of his ruled white pad, and then tear it off and insert it into a fresh light-brown manila folder which he labeled and then inserted into one of his file drawers. In an article published after his death, Beverly [Beverly Bell, one of his associates at Goldman] described 6000 files he left behind, now archived at MIT."[10]

By 1986, the three of them had created a bond option model that seemed to work. One of the things that Derman was able to bring to the collaboration was the skill at computer programming that he had acquired at Bell Labs. To make such a model useful to traders there had to be a computer program that they could use to make very rapid estimates of the price of the bond options they were selling. They created such a program. But because of Black's fastidiousness it took almost four years before a paper that he considered satisfactory was published. Because of its simplicity and accessibility to traders, the BDT model, as it became known, was widely used in the industry. In 1994, Black developed the cancer that would take his life a year later. He was as frank and outspoken about his impending death as he had been about everything else. During his illness he had written a paper that he called "Interest Rates as Options." Knowing the seriousness of his illness, in May of 1995, he wrote to the editor of the *Financial Analysts Journal*, to whom he had submitted his paper, "I would like to publish this, though I may not be around to make any changes the referee may suggest. If I am not, and it seems roughly acceptable, could you publish it as is with a note explaining the circumstances."[11] The paper was published without his completed revisions after he had died in August, at the age of 57.

In 1988, after he had been at Goldman for a relatively short time, Derman decided that he needed a change of scene. He interviewed at Salomon, where eventually he took a job for a very unhappy year, after which he returned to Goldman. One of the groups at Salomon that he interviewed with was one that had been handpicked by John Meriwether. This group had the reputation of being the savviest derivative traders on Wall Street. Derman had lunch with them, and in a very polite way they grilled him. One of the questions they asked concerned "Asian" versus European options. He didn't know the answer. But this was a bit of a trick question since these options, which involve certain kinds of averaging procedures, are only called "Asian" because a couple of traders from Bankers Trust happened to be in Tokyo in 1987, when they thought them up. In any event, Derman did not get the

---

[10] ML p. 145.
[11] ML p. 168.

job. This was probably fortunate because ten years later this group precipitated a crisis in the financial markets that could have led to a total economic meltdown. It didn't because, under the auspices of the Federal Reserve, a consortium of investment banks spent billions to salvage the situation.

In reading about this (Roger Lowenstein's book *When Genius Failed*[12] is an excellent source) I was struck by the difference between this financial scandal and ones that we are today more familiar with—Enron, Global Crossing, and the rest. In the first place, there is the scale. While the Enron scandal, for example, was a financial disaster for a large number of people, it was never a threat to the system as a whole. (The present mortgage crisis which involves the mis-rating of sub-prime loans, may well be.) In the second place, there was the matter of intent. Many of the people involved in Enron-like scandals often end up by going to jail. They participated in criminal activities. For the participants in Long Term Capital Management-LTCM—which is what Meriwether's hedge fund was called—there was no criminal intent. I am not even sure how interested they were in money, except as a measure of how smart they were. If they had been asked the question that Bernie Cornfeld used to ask prospective employees of International Overseas Services, another investment disaster, "Do you sincerely want to be rich?" (meaning, would you sell your sister), I think they might have had some difficulty answering. Not long after Meriwether started his fund in 1993, he successfully recruited both Merton and Scholes. In his 1997 Nobel Prize autobiography, written a year before the final catastrophe, Merton was euphoric about the fund. This is what he wrote:

"This small group of founding principals, together with a few key early employees, put together and tested the financial, telecommunication and computer technologies, hired the strategists and operations people to run them, designed the organizational structure of the business, executed the complex contractual arrangements with investors and counterparties, found and outfitted physical quarters in both the United States and London, and helped to raise over $1 billion from investors. The design and development efforts along each of these dimensions attempted to marry the best of finance theory with the best of finance practice. It all came together in February 1994 when the firm began active business. Today LTCM has 180 employees, a third office in Tokyo, and its capital has grown considerably."

Merton goes on, "It was deliciously intense and exciting to have been part of creating LTCM. For making this possible, I will never be able to adequately express my indebtedness to my extraordinary talented LTCM colleagues."

The distinctive LTCM experience from the beginning to the present characterizes the theme of productive interaction of finance theory and finance practice. Indeed, in a twist on the more familiar version of that theme, the major investment magazine, *Institutional Investor*, characterized the remarkable collection of people at LTCM as "The best financial faculty in the world."[13]

---

[12] Random House, New York, 2000. This will be referred to as "Genius."

[13] This autobiography as well as the one written by Scholes can be found on the website www. nobel.se/economics/laureates/1997.

One wonders what the Nobel committee made of this, to say nothing of what they and the editors of *Institutional Investor* made of it the following year when "The best financial faculty in the world" came close to wrecking the entire world's financial infrastructure.

The "dean" of this dream faculty, John Meriwether, was born on the South Side of Chicago in 1947. He came from a middle-class Catholic family. His father was an accountant and his mother worked for the Board of Education. Meriwether was educated in very strict parochial schools. He was a good student, but not exceptional, except perhaps in mathematics. If it hadn't been for golf it is not clear where he would have ended up. He was an extremely good golfer. While working as a caddy at the Flossmor Country Club, which had some very wealthy members, he was selected for a college scholarship that was awarded only to caddies. He chose to attend Northwestern University. After graduation, he first taught mathematics in high school for a year and then went to the University of Chicago to study business. One of his classmates was Jon Corzine, now the governor of New Jersey, who, when he was the CEO of Goldman, became involved in the LTCM denouement. In 1973, Meriwether went to work at Salomon. This was just before the explosion in derivative trading. Indeed, in 1977, Meriwether began assembling the Arbitrage Group at Salomon, the people that Derman had his unsuccessful interview with, and who later formed the core of LTCM.

In assembling his group Meriwether sought people from all over who were smarter than anyone else, smarter even than he was. He had no complexes about this and no problem in seeking misfits, nerds, and geeks from academia so long as they were brilliant. These people, who were characterized by another Salomon trader as "freaks" and "a bunch of guys who would be playing with their slide rules at Bell Labs"[14] if they hadn't been tapped by Meriwether, loved the financial engineering models. They win when the market is a universe of inefficiency, a garden salad of incorrectly priced derivatives that they could gobble up and then wait for what they were certain would be a return to efficiency, at which point they would make a killing. The key to everything was the assumption that the market would behave rationally. This continuity of behavior was one of the assumptions, for example, that goes into deriving the Black-Scholes formula. If in the Brownian motion, for example, the drunkard suddenly falls down a manhole, all bets are off.

For several years Meriwether's group thrived and Meriwether became richer and richer. One of the things he did was to invest in thoroughbred horses, but overall he was still the rather unassuming parochial school boy. There was nothing in his group, then or later when they were LTCM, which remotely resembled the sort of partying that Bernie Cornfeld was famous for in Geneva. After attending one of his parties John Kenneth Galbraith quipped, "Where are the customer's girls?", echoing something that had been said many years earlier about yachts. Meriwether's very tight-knit group played "liars poker" or golf. What they did not do was to explain to people outside the group anything about their trading. This was one of the characteristics of LTCM. Banks and brokerage houses put in millions and

---

[14]Genius, p. 13.

millions without having any real idea of how the money was being invested. Hedge funds were relatively new and almost unregulated. They still are. All that mattered to these investors was that out of the black box vast returns kept appearing.

Here is a little analogy that may be useful in understanding the denouement. I propose a scheme that guarantees that I will win $1,000 at the roulette wheel in Monte Carlo. I will bet $1,000 on red. If it comes up red I will collect my money. If it comes up black I will bet $2,000 on the next turn of the wheel. Again if it comes up black I will double my bet and so on. Unless the wheel is crooked it must sooner or later come up red, so that I will win my money back. But, even if it is an honest wheel, there can be fluctuations. If, for example, it comes up black ten times in a row my next bet will run into the millions. Perhaps I can persuade a bank to lend me the money to keep going. This is called leverage. If not, Lord Keynes maxim that was quoted at the beginning of this section will come true. Once I start the game I can't stop unless either I hit red or am prepared to pay off the last bet. Seeing this, the bank may want its money back or the casino may decide that I have reached the limit at which they will take a bet from me. Either of these situations is potentially catastrophic and both of them in a manner of speaking came to apply to LTCM. It is quite right to think that in the long run black must come up, or the markets must return to rationality. But there is another Keynes maxim that says simply that in the long run we are all dead.

In late summer of 1998, the fund had $3.6 billion in capital. This unknown enterprise in Greenwich, Connecticut, where this relatively small group was located, was a larger financial enterprise than any of the major brokerage houses such as Merrill or Goldman. But during a five-week period in August and September they lost it all. They were wiped out. The "faculty" had personal losses of $1.9 billion. This is not the place to adumbrate this catastrophe in detail. Lowenstein's book gives a detailed account. However, I would like to give the flavor. In this, our toy model will again be helpful.

Recall that the model is a toy because there are only two outcomes for the stock—$120 or $80. The difference between these numbers, $40, or the spread, is a measure of the risk. The risk is reflected in the amount that the hedge will cost us. In this example it was $10, the price of the option. But suppose we widen the spread and take $140 and $60 so the spread is now $80. If we do the algebra here we will find that the cost of the hedge is now $13.33. The risk is higher. In the real world we do not have two states, so we must have some theory for the future volatility. This is what LTCM thought it had. So its traders looked for stocks where in their view the volatility had been overestimated. Japan was a good source, which is why LTCM opened an office in Tokyo. Owners of these stocks were ready to pay LTCM a premium to create a hedge. The assumption was that the market would behave rationally so that in the course of time the volatility—the spread—would relax to the predicted value. It *had* to, just as the roulette wheel had to come up red. But there was an additional element. LTCM was not playing the game with its own money. It was playing it with borrowed money, leverage. It took advantage of the very loose regulation that there was for hedge funds. Alan Greenspan, for example, thought that they should not be regulated at all. Market forces would take care of things—an attitude that still persists. Thus there was no objection to LTCM's circumvention of the leverage regulations for buying stocks. It did so by giving borrowed money to banks, which would set up accounts ("swaps") that mirrored in terms of profits and loss the stock that LTCM wanted to buy. There was no limit on this, and banks were just shoveling money at the firm.

But by the spring of 1998, it all became unglued. Instead of narrowing, the spreads became wider. This meant that LTCM had lost its bet on the option cost. It has also begun to make investments directly in stocks, and these were also losing money. Markets around the world were sinking. In August, Russia defaulted on its external debts, causing further chaos in the financial markets. Everyone was looking for liquidity, and with its huge positions LTCM could not unload. To add to everything, LTCM had an arrangement with Bear Sterns, who acted as LTCM's broker of record. They actually carried out the transactions. The condition was that they would stop doing this if the reserve they held from LTCM— "cash in the box" —fell below $500 million. This was money based on LTCM's assets, which were rapidly melting away. If this reserve fell below the $500 million Bear Sterns would stop executing trades— the roulette wheel would stop—and LTCM would be out of business.

Meriwether tried without success to borrow money from anyone he knew, including Warren Buffet and George Soros, to keep the thing afloat. But by the middle of September it was clear that, without outside help, the company would collapse and, because of its intertwining relationships with banks and brokerages both here and abroad, the market might itself collapse. By end of September, in a much criticized move, the Federal Reserve orchestrated a rescue in which fourteen banks provided $3.65 billion to take over the fund. Long Term Capital Management was gone.[15]

---

[15]That this liquidity crunch could cause a disaster should have been predictable since a very similar thing occurred in the fall of 1987. At this time so-called "portfolio insurance" was rather widespread. The idea was to insure against falls in the stock market. The device for doing this was a so-called "put." A put is a contract taken out when the stock is at its present price. It allows one to sell the stock at some specified future time at the price that prevailed when one bought the put. If the stock falls one can sell it for more than its current value. If the stock rises one does not exercise the put and is only out its cost. For various reasons, actual put contracts were not convenient for this activity. Instead one used the Black Scholes-Merton mathematics to replicate the put with a mixture of bonds and stocks. The mathematics tells one how to continually re-adjust the mixture. The stocks are sold "short." This means that you borrow the shares, sell them at the prevailing price, and then if the shares fall you can repurchase them at the lower price to return to the lender. The mathematics tells you that as the stocks in the portfolio fall you must buy larger and larger quantities of the hedging shorts to cover them. It was assumed erroneously that selling these shorted stocks would somehow have no effect on the market. In fact it simply increased the market decline which meant more and more shorts had to be sold, accelerating the decline. This was the source of the disaster. Incidentally, Professor Zvi Bodie of Boston University proposed a nice application of these ideas. Mutual funds and other purveyors of stock often tell us that if we hold onto a stock long enough this will mitigate the risks. Professor Bodie asked what would happen if we insured such a stock with a put. If the mutual funds were right the cost of the insurance should diminish in time. We can analyze this with the Black-Scholes mathematics if we replicate the put by a call option. What happens is that the cost of the insurance increases monotonically until it reaches the cost of the stock itself, showing that holding risky stocks makes them more risky over time.

Despite their losses the partners came out of this debacle as wealthy men. It does not seem to have destroyed their professional lives. Merton is a professor at Harvard Business School. Scholes is a partner in a management firm in which Robert Bass has the controlling interest. Meriwether hardly missed a beat. In 1999 he started a new firm called JWM Partners in Greenwich. At last reckoning this firm had about $2.6 billion under management. The roster of its associates turns up several of the same names that are familiar from LTCM. In its heyday there was always a split at LTCM between the conservatives, such as Merton and Scholes, who wanted to stick with their models, and the traders, who were simply interested in speculation. Meriwether chose the traders. During the whole debacle he said next to nothing except to apologize to some of the bankers. The impression one has is that these people were much more worried about prestige than money. In this respect they needn't have worried. Financial engineering is thriving. Derman's courses at Columbia are an example. So are the hedge funds. They used to be a tool of the very rich. But now even pension funds invest in them with predictable negative results. A recent example is the San Diego County retirement fund, which invested $175 million with Amaranth Advisors in 2005. Amaranth managed to lose most of it. The retirement fund is suing. Despite this and other examples there is still no meaningful supervision of these funds, which operate largely in secret—secret even from their investors. I can't help thinking of what Einstein replied when he was asked what he would have felt if experiments had failed to confirm his theory of gravitation: "Then I would have felt sorry for the dear Lord—the theory is correct" "*die Theorie stimmt doch.*"

Part II
Scientists

Chapter 4
Heisenberg in Poland

In 1980, Elisabeth Heisenberg published a memoir in which she attempted to explain the complex actions of her late husband during the war. (He had died in 1976.) It was translated into English under the title "Inner Exile."[16] "Inner exile," was, she says, a status that her husband chose for himself. He had decided not to emigrate and he had no wish to become a martyr, so he remained in Germany making whatever compromises were necessary to survive and work.

The picture Elisabeth paints of her husband is, as one might imagine, unfailingly flattering. It is, however, a portrait in which things are often left out or distorted. One could cite any number of examples, but we will focus here on what she has to say about Heisenberg's war-time visits to several countries occupied by the Germans: "A further duty Heisenberg felt bound to and thought to be important was to give scientific lectures as often as possible—either at native or foreign universities, but especially at the universities of the occupied areas—so as not to lose contact with his harried colleagues as well as to demonstrate that a different, better Germany existed than the Nazi Germany that had won the upper hand to such a terrifying degree."[17]

A nice picture, but is it true? Does it apply, for example, to Heisenberg's visit to occupied Poland, which took place in December of 1943, eight months after the Germans had liquidated the ghettos of Warsaw and Cracow? What did Heisenberg know about the extermination of the Polish Jews? Did he on his visit in fact restore contact with his "harried" Polish colleagues?[18]

---

[16] *Inner Exile*, by Elisabeth Heisenberg, Birkhäuser, Boston, 1984.

[17] Heisenberg, op. cit., p. 96.

[18] This visit has been discussed in various degrees of completeness by David Cassidy, *Uncertainty*, W. H. Freeman, New York, 1992, Thomas Powers, *Heisenberg's War*, Alfred Knopf, New York, 1993, Mark Walker, *Nazi Science*, Plenum, New York, 1995, and *German National Socialism and the Quest for Nuclear Power, 1939–1949*, Cambridge University Press, Cambridge, 1989. I am very grateful to these authors for many helpful communications.

J. Bernstein, *Physicists on Wall Street and Other Essays on Science Society*,
© Springer Science + Business Media, LLC 2008

The route that ultimately led to Heisenberg's 1943 visit to Cracow began at the time of the First World War, in Munich. Heisenberg was then enrolled in the Maximillians—"Max"—Gymnasium, as was his older brother Erwin. For a period, Erwin had a classmate named Hans Frank.[19] Both boys had been born in 1900, while Heisenberg was born a year later. Hans Frank and Erwin Heisenberg graduated from the gymnasium in 1918, after which Frank served in the army infantry for a couple of years. The precise amount of contact the Heisenbergs had with Hans Frank is not clear. There appears to be no mention of Frank in any of the Heisenberg correspondence so far discovered, at least until 1943.[20] Werner Heisenberg became a *Pfadfinder*— "pathfinder." The Pfadfinders were the German boy scout movement. Somewhat later he joined the *Neupfadfinder*, the "new pathfinders," a group that added Teutonic romanticism to the usual hiking and camping. About the same time Hans Frank also joined the Neupfadfinders. For Frank, and others, this Teutonic mystic romanticism led in due course to embracing National Socialism.

Heisenberg, neither then nor ever, was a member of the Nazi Party, nor any of its offshoots. Things were very different with Frank.[21] After his military service he began the study of law at the universities of Kiel and Munich and almost immediately joined a right-wing paramilitary group called the Epp Freikorps. By 1923, he had become a Storm Trooper and a member of the Nazi Party, taking part in Hitler's failed Beer-Hall putsch in Munich. He had a love of Hitler that bordered on the erotic. He soon began moving up in the party's ranks by defending various Nazis in libel suits. At one point Hitler asked him to examine the Führer's family tree to see if any Jews were lurking among the branches. In 1933, he was appointed minister of justice for the state of Bavaria and, soon after, minister without portfolio. His general task was to create a legal façade behind which Hitler's regime could operate with the fiction of a legal system. One of the items that he later cited in his defense was the role he played in the so-called "Night of the Long Knives." In June of 1934, Hitler organized a massacre of people he felt were a challenge to his power, among them Ernst Röhm, who was the leader of the Storm Troopers. Frank claimed that by intervening with Hitler personally he reduced the number of people executed from 110 to 20. Hitler proposed that Röhm should be allowed to commit suicide but, when he refused, he was shot.

---

[19] David Cassidy has studied the records of the Max Gymnasium. He informs me that in the years 1911–1914 Erwin Heisenberg and Hans Frank were in the same class but different sections. However, in 1914 there was only one section in which Frank and Erwin Heisenberg were both enrolled.

[20] I am grateful to Helmut Rechenberg of the Heisenberg Archive in Munich for pointing this out and also for supplying an account of the correspondence that led to Heisenberg's visit.

[21] A scathing portrait of his mother and father is given by Niklas Frank in his book *In the Shadow of the Reich*, with Arthur S. Wensinger, Carol Clew Hoey (Translator), Jonathan B. Segal (Editor). Knopf, New York, 1991. Frank, who collaborated in a play about this material, was born in 1939, but his memory of both wartime and the immediate post-war experiences is very vivid. There is as yet no biography of Frank. The following websites with their links may be useful. http://www.dhm.de/lemo/htm/biografen/Frank Hans/ and http://www.wikipedia.org/wiki/Hans_Frank.

The Germans invaded Poland on September 1, 1939. A month later, Hitler appointed Hans Frank as the governor general of Poland, with headquarters in Cracow. Indeed, he settled with his wife and family, including his infant son Niklas, in the Wawel Castle in Cracow, whose name he changed to the German *Krakauer Burg*.

It is important to understand for what follows that this Germanization of a name was not an isolated whim but was part of a systematic effort to reduce Poland to a colony without a culture, an appendage to the Reich designed only to serve its needs. The intent was made absolutely clear. Frank stated it himself: "What we recognize in Poland to be the elite must be liquidated."[22] Poland, he said, was to "become a society of peasants and workers," with no "cultured class." This meant that as far as the Poles were concerned higher education was to stop; along with Polish theater and literature. The language itself was to be obliterated. No radios were allowed, and all news came from loudspeakers that belonged to the Nazi authorities. Undesirable books were banned even if they had not been written by Jews. Jews, of course, were herded into ghettos, where they were readily available for shipment to extermination camps.

Cracow surrendered to the Germans on September 6, 1939. The university was soon to open for its fall term. On November 3 Bruno Müller, the local Gestapo chief, ordered the rector of the University, Professor Lehr-Splawinski, to call a faculty meeting for November 6 at noon.[23] He assumed that Muller was going to discuss—it was still early days—the sort of higher education that would be encouraged under the occupation. Some 155 of the invitees came, including various other university employees. It was a trap. They were all arrested on the spot by the SS. The SS also rounded up anyone else who happened to be in the building—a total of 183. This operation became known as the *Sonderaktion Krakau*, Special Action-Cracow. Its anniversary is still acknowledged at the university, where a ceremony is held annually in the room in which they had been arrested. After a few days in a local jail, they were shipped to Breslau and then to the concentration camp at Sachsenhausen-Oranienburg.

News of this got out to other European scientists and, in part because of their protests, on February 8, 101 scientists—mainly men over age 40—were released.[24] Twelve had already died in the camp, and the rest were sent to other concentration camps or remained in Sachsenhausen. A few more were subsequently released, while the remainder, including all the Jews, were killed. The university was now

---

[22] This quotation and the one that follows can be found in a fascinating essay called "The University of Cracow Library under Nazi Occupation: 1939–1945" by Mark Sroka, *Libraries and Culture*. Vol. 34, Winter 1999. Sroka is primarily concerned with the fate of the Polish libraries, but he also discusses the general cultural life.

[23] For information on this event and many other aspects of this history, I am greatly indebted to Kryzstof Fialkowski, who is a theoretical physicist on the faculty of Jagellonian University in Cracow. He discussed Heisenberg's visit with colleagues who have recollections and he also searched newspaper archives and other historical sources.

[24] For a discussion of this see, for example, *Germany Turns Eastward* by Michael Burleigh, Cambridge University Press, Cambridge, 1988, pp. 253–254. When a letter was circulated in Germany protesting, the only physicist to sign it was Max von Laue.

closed. Any higher education had to be carried out clandestinely at peril to teacher and student alike.[25]

Frank and the SS acted sometimes at cross purposes, sometimes in concert. There was a constant power struggle between them. Frank was himself a crude and brutal anti-Semite. During this whole period he maintained a journal. Ultimately the entire journal came to 43 volumes.[26] It included records of his speeches, one of which was addressed to his cabinet on December 16, 1941. Here is some of what he said,

"As far as Jews are concerned, I want to tell you quite frankly that they must be done away with in one way or another. The Führer said once: 'Should united Jewry again succeed in provoking a world war, the blood of not only the nations which have been forced into war by them, will be shed, but the Jew will have found his end in Europe' I know that many of the measures carried out against the Jews in the Reich at present are being criticized. [One wonders by whom.] It is being tried intentionally, as is obvious from the reports on the morale, to talk about cruelty, harshness, etc. Before I continue, I want to beg you to agree with me on the following formula: We will principally have pity on the German people only, and nobody else in the whole world. The others, too, had no pity on us. As an old National-Socialist, I must say: This war would only be a partial success if the whole lot of Jewry would survive it, while we would have shed our best blood in order to save Europe. My attitude towards the Jews will, therefore, be based only on the expectation that they must disappear. They must be done away with. I have entered negotiations to have them deported to the East. A great discussion concerning that question will take place in Berlin in January, to which I am going to delegate the State Secretary Dr. Bühler. That discussion is to take place in the Reich Security Main Office with SS-Lt. General Heydrich. A great Jewish migration will begin in any case."

Lest there be any confusion about what Frank meant by "migration" he goes on,

"But what should be done with the Jews? Do you think they will be settled down in the Ostland, in villages? This is what we were told in Berlin: Why all this bother? We can do nothing with them either in the Ostland nor in the Reich kommissariat. So liquidate them yourself."[27] But it was not only the Jews. Frank decided that the German war machine needed manpower, and by 1940, he was making arrangements to export slave labor to Germany. By August of 1942, he had supplied 800,000 Polish workers for the Reich. Of the Jews, on January 25, 1944, Frank estimated that of the original 2.5–3.5 million in his territory, only 100,000 were left.

---

[25] I know a Polish physicist, Jacques Prentki, who received his education this way. Prenkti was himself arrested in a random operation in Warsaw but managed to escape the box car in which he had been placed in which he was being shipped to an extermination camp. He is not Jewish. Professor Fialkowski informs me that on the day of Sonderaktion his mother, who was a law student, was in the library across the street. She was with a friend who went to see what was happening and did not return for six months. He was later killed in the 1944 Warsaw uprising.

[26] There is some disagreement about the number of volumes. At Nuremberg Frank said 43, but only 38 were actually found. The National Archives and Records Administration has these on microfilm. Part of the copy I studied was very dark and not easy to read.

[27] This quote can be found on the website http://fcit.coedu.usf.edu/holocaust/resource/document/DocFrank.htm. It is translated from the German, and I have quoted this translation.

With this background we can return to Heisenberg. The first thing to remark was that there is no credible evidence that Heisenberg was ever an anti-Semite. His postdoctoral employer in Göttingen was Max Born, who was a Jew. In one of his finest actions, when Heisenberg won the Nobel Prize in Physics for 1932—it was actually not awarded until 1933—feeling that Born should have shared it, he crossed the border into Switzerland so that he could mail an uncensored letter in which he expressed his regrets to Born. He was nominated for the prize by both Einstein and Bohr. His closest collaborator was Wolfgang Pauli, who used to tease Heisenberg about his Pfadfinder connections. There were many people such as Rudolf Peierls (later Sir Rudolf Peierls) who came as students to work with him. All these people had Jewish ancestry.

I said earlier that there is also no evidence that Heisenberg was ever a Nazi. Until the war, he seemed largely indifferent to politics, although he was a thoroughgoing German patriot and nationalist. There is evidence that not only did Heisenberg want the Germans to win the war, but that he felt that the invasion of Poland was a good thing. He also, as the following will show, was a person with an almost pathological lack of understanding of the feelings of other people, at least in some circumstances.

As is well known, Heisenberg made a visit to Copenhagen in September of 1941. We now know that during this visit, he spent three evenings with Bohr and his wife at Bohr's house.[28] But besides this he made visits to Bohr's institute, where he spoke with some of the physicists. Among them was Stefan Rozental, who had been born in Poland. Rozental retained a vivid memory of this encounter, which he described in a letter to the British historian Margaret Gowing.[29] Rozental wrote:

"He [Heisenberg] stressed how important it was that Germany should win the war. To Christan Moller [a well known theorist at the institute], for instance, he said that the occupation of Denmark, Norway, Belgium, and Holland was a sad thing but as regards the countries in East Europe it was a good development because these countries were not able to govern themselves. Moller's answer was that so far we have only learned that it is Germany that cannot govern itself."

We do not know how much Heisenberg knew of his school friend's activities in Poland, but we do know that early on he knew that Jews were being slaughtered there. We know this from none other than Elisabeth Heisenberg herself. In her book she addresses the question of why good Germans like herself continued to deny the reality of what was happening around them. She cites the following example:

"I can still see my father standing in front of me. He was a man with a venerable and law-abiding outlook, who actually went into a rage when Heisenberg once showed him a report he had received from a colleague at the institute who had been a witness to the first cynical mass executions of Jews in Poland. My father lost all self-control and started to shout at us: So this is what it has come to, you believe things like this! This is what you get from listening to foreign broadcasts all the

---

[28]This recently came to light in a letter that Heisenberg wrote to his wife from Copenhagen. The letter in English and German can be found at http://werner-heisenberg.unh.edu

[29]The letter is quoted in Powers, op. cit., p. 121.

time. Germans cannot do things like this, it is impossible! He was not a Nazi; he had prematurely retired from his position following the National Socialist takeover."[30]

One would give much to know what this report said and who wrote it. She does not tell us.[31] But what strikes one is that she apparently sees no connection between this report and Heisenberg's subsequent visit to Poland. She doesn't seem to realize that what she is saying, albeit indirectly, is that when Heisenberg visited Poland he knew in advance of the slaughter of the Jews. The question which begs for an answer then is, why did he go? Why didn't he, by refusing this visit—which he could have done by claiming, for example, that he was fully occupied doing research related to the war—at least make some small gesture of protest, an inner exile?

I have not been able to discover whether, and to what extent, if any, Heisenberg kept up his contact with Hans Frank between the time when they were in the Pfadfinders and when he received his first invitation to visit Poland in May of 1941.[32] Perhaps we will learn more when, and if, the Heisenberg family publishes the thousands of private letters (an historical treasure) that they have kept to themselves. The invitation to visit Poland was not signed by Hans Frank but by a man named Wilhelm Coblitz, who was the director of something called the *Institut für Deutsche Ostarbeit*—The Institute for German Work in the East. This group had been formed in the spring of 1940 by Frank. It was devoted to studies in aid of the colonization of the eastern countries.[33] The astronomy and mathematics section used Russian forced laborers. Much of the research was devoted to the Jewish question and to racial matters in general. This invitation was issued on behalf of this institute. Heisenberg was quite willing to accept it, but he was not given permission to make the trip.

Here we must back up a little. During a brief period Heisenberg was himself under suspicion. He had even been called a "White Jew" because of his association with Jewish scientists and his unwillingness to accept and teach an absurd Aryan physics that had become part of the Nazi ideology. Finally, the matter, which could have become very serious, was sorted out by an intervention from Heinrich Himmler. It had been arranged through the two men's mothers, who knew each other. An agreement was reached that Heisenberg could use the physics provided that he did not discuss its non-Aryan origins. Even after this had happened he was not able to get permission to travel.

---

[30] Elisabeth Heisenberg, op. cit., p. 49.

[31] Rechenberg believes that it was the physicist Karl Wirtz and that the report was in 1942. Wirtz was one of the ten German scientists detained at Farm Hall near Cambridge. These conversations were recorded—see my *Hitler's Uranium Club*, Copernicus, New York, 2001. In one of them Wirtz says, "We have done things which are unique in the world. We went to Poland and not only murdered Jews, but for instance, the SS drove up to a girls' school, fetched out the top class, and shot them simply because the girls were high school girls, and the intelligentsia were to be wiped out." p. 98, Bernstein, op. cit. This does not appear to be the incident described by Elisabeth Heisenberg. There is no reason to assume that these girls were Jewish.

[32] Mark Walker's book *Nazi Science*, op. cit., has been very helpful to me with these details.

[33] For a full discussion of this see *Burleigh*, op. cit.

This changed in the fall of 1941. Heisenberg had a student and protégé named Carl Friedrich von Weizsäcker, whose father Ernst was State Secretary—the highest form of civil servant—and was in a position to alter Heisenberg's travel status. Indeed, Heisenberg and von Weizsäcker got permission to attend a conference of astronomers in Copenhagen. This was the occasion on which they both visited Bohr in his home. It appears as if the German authorities regarded the visit to Copenhagen as a success, since subsequently Heisenberg readily got permission to travel to places such as Holland and Switzerland and, ultimately, to Poland.

Incidentally, there was an odd trace of the 1941 aborted visit. In the German language Cracow daily—the *Krakauer Zeitung*—in January of 1942 there appeared in two successive issues a lecture by Heisenberg entitled "Unity of the Scientific Worldview." It turned out that it had been given at Leipzig University on November 26, 1941. It was very likely the lecture Heisenberg would have given if he had gotten permission to come to Cracow.

Coblitz renewed his attempt to get Heisenberg to visit Poland in May of 1943. This time he wrote in the name of Frank, as well as himself, to urge Heisenberg's visit. In subsequent letters he conveyed Frank's "*besten Grüse*"—best greetings— and Heisenberg responded in kind.[34] Coblitz said that Frank would personally attend the lecture that Heisenberg was scheduled to give. There was then a hiatus because Frank's summer vacation plans had not been fixed. But on September 29 Coblitz writes, "*Der Herr Generaldirektor läst Sie und Ihre Frau einladen, seine Gäste auf Schloss Wartenberg, näher bei Krakau zu sein.*"[35] "The *Herr Generaldirektor*, Dr. Frank invites you and your wife to be his guests at the Wartenberg Castle, near Cracow." The Wartenberg Castle is a villa that was built between the wars within sight of the large Wawel Castle. Ironically, it now belongs to the university and is used for conferences on Polish culture. The name "Wartenberg"—"Observatory Mountain"—was a German invention used by Frank and his associates. The villa was known locally either as the "Szyszko-Bohuz villa," after the architect who built and owned it, or the "Przegorzaly villa," after the town above which it is located. In the summer of 1943, Frank had "donated" it, after his visit, to Himmler for use by the SS. But he continued to occupy it. Heisenberg responded that while he accepted the invitation, his wife was unable to go because of her domestic responsibilities.

For several reasons, we do not have as complete a record of Heisenberg's December 1943 visit to Cracow as we do for his visit to Copenhagen, for example, although even here the record is still incomplete. One reason is that during his visit Heisenberg saw none of the Polish physicists. This is perhaps not surprising since, in some sense, university professors were regarded as outlaws by the regime. How this fact struck Heisenberg we do not know because no report of the visit written by him has been found. Also there is no photograph that I have been

---

[34] I am grateful to Mark Walker for a file of these letters and to Helmut Rechenberg for permission to quote from them.

[35] The quotation was supplied by Rechenberg.

able to locate. We do know that he stayed in Frank's castle. The previous June, Himmler had stayed with Frank in the Cracow castle. What Heisenberg thought of the artwork is, as far as I know, not recorded. Frank's castles were furnished with masterpieces stolen from the Poles—some from museums, some from cathedrals.[36] Frank estimated that 90 percent of the valuable art in his territory had been "safeguarded." Hitler ended up with 30 Dürers. Frank furnished his domiciles with works of artists such as Leonardo da Vinci, Raphael, and Rembrandt. As the war was ending, many of these were shipped to Germany. Frank explained this to the GI's who eventually captured him by saying, "I took along certain objects of art so that they would not be plundered in my absence."[37] Some of the art was later restored and some simply disappeared. One wonders at Heisenberg's feelings when he discovered that he was staying in an art museum. He must have known where these treasures came from.

Heisenberg gave a lecture. Although I do not have a copy of the text, I have been informed[38] that no Poles, although they tried, were allowed to attend, only Germans. The Poles were turned away at the door. In the December 18 issue of the *Krakauer Zeitung* the following article appeared under the title "The Smallest Building Blocks of Matter."[39]

"Prof. Dr. Werner Heisenberg, Director of the Kaiser-Wilhelm-Institut für Physik, Berlin-Dahlem, lectured to a large audience of interested listeners in the great lecture hall of the Institut für Deutsche Ostarbeit about the central problems of scientific progress: contemporary aims of research in physics. The lecturer presented the development of modern atomic physics from their beginnings at the end of XIX-th century. At the start of this development there was the discovery of Roentgen rays and Planck's theory of quanta. It was finalized about 15 years ago by the so-called quantum mechanics (Heisenberg). Since about ten years the main line of research in atomic physics became the investigation of atomic nuclei. By the use of high voltage devices and other high technology means it became possible to transform the atomic nuclei, and thus to fulfill the old program of the alchemists: the transmutation of chemical elements. However, the ideal laboratory, in which the atomic transformations occur at highest energies, was presented to us by nature in the form of cosmic rays. The sources of this strange radiation in space are unknown. However, the effects of this radiation are being investigated by physicists and provide

---

[36] For an account of this plunder see *Art as Politics in the Third Reich* by Jonathan Petropoulos, University of North Carolina Press, Chapel Hill, 1996, and *The Rape of Europa*, by Lynn H. Nicholas, Vintage, New York, 1995.

[37] Niklas Frank, op. cit., p. 309.

[38] By Professor Fialkowski, who asked those of his colleagues who knew physicists who were in Cracow at the time. There were a few who had actually been there and had tried to attend the lecture.

[39] I am very grateful to Professor Fialkowski for finding this article in the library. The reference is Krakauer Zeitung, 1943, nr, 302, December 18. He also sent me the German original, from which this is a translation.

us with most interesting information about the nature of the smallest building blocks of matter.

After the enthusiastically received lecture, Governor-General Dr. Frank spoke personally as the president of Institut für Deutsche Ostarbeit and praised the work of the lecturer, who is among the most eminent personalities of the internationally recognized German science. Heisenberg, a Nobel Prize winner at the age of thirty, belongs to the list of great German physicists, whose investigations in theoretical physics led to landmark discoveries."

The closest Heisenberg ever came to explaining his visit to Cracow was in an interview he gave to David Irving in 1965. What Irving reports Heisenberg as saying is this:

"Here in Munich I was in school with some people who later became great Nazis, among them the Herr Generalgoveneur of Poland—Frank. Frank was in the school class of my brother, and so naturally he knew us and *dutzten* us. [This is not really translatable, since in English we do not have the equivalent of the formal "you"—*Sie*—and the familiar "you"—*Du*. The implication is that the friendship was close enough so that the familiar was used.] I had completely lost sight of him and thought, OK, I will have nothing further to do with him. But then around September of '43, if I remember correctly, he wrote that I should nevertheless come to Cracow, and give a scientific lecture there. I felt, this is stupid, what am I doing there in Cracow; Frank does not concern me anyway. But he wrote in such a friendly way: my dear friend! Can you not...so that I wrote: Dear Frank! Well, I have so many other things to do here, unfortunately it is impossible for me to come. But then he sent me yet another letter, and was so pressing, and with implications which did not sound so pleasant, so I thought I do not really need to make an enemy. OK, I will give the lecture in Cracow. So in December 1943, if I remember well, I went to Cracow where first I was his guest in his castle, then I gave a lecture on the innocent theme of quantum theory, or something like it."[40]

What is one to make of this? Firstly it must be noted that no trace of any letter from Frank to Heisenberg has been found. The only correspondence that is known is between Heisenberg and Coblitz acting on Frank's behalf, which began in 1941. I have quoted from some of it. Nowhere does Heisenberg show any reluctance to go to Cracow, and nowhere is there any suggestion of a parallel correspondence with Frank. On the contrary, the letters always convey anything personal through Coblitz as the intermediary. It is as if in this interview Heisenberg has created a fantasy of what, looking back, he would like to have happened, and how he would like for it to be perceived. Like so much else that involves Heisenberg we end up with an enigma. What did Heisenberg really think about

---

[40]Niels Bohr Library, AIP, M140, 31526–31567. In the translation above I have left out the allusions to nuclear weapons. My problem with them is, considering the unbelievability of the rest of the letter, what are we to believe about this?

his visit to Poland? How could he possibly justify it? Was he blind to what was going on in Poland all around him? With Frank there was no enigma. In October of 1945 he went on trial at Nuremberg.[41] He was found guilty and on October 16, 1946, he was hanged.

---

[41]In his testimony at Nuremberg, Frank stated that he did not arrive in Cracow until a few days after the Sonderaktion. He then tells us he devoted himself to getting the imprisoned faculty released. However, it was pointed out to him that in his journal he said that these professors should be returned to Poland either for liquidation or imprisonment. To this he responded that he had written that "to hoodwink my enemies." Frank also claimed that he encouraged higher education in Poland under the occupation, something which certainly would come as a surprise to the people who lived under it. This testimony can be found on the site http://www.law.umke.edu/faculty/projects/ftrials/nuremberg/ franktest.html. In his diary Frank records his comment on the Sonderaktion:

"We cannot burden the Reich concentration camps with our affairs. The trouble we had with the Cracow professors was awful. Had we dealt with the matter here it would have taken a different course. I should therefore like to request you urgently not to deport any more people to the concentration camps in the Reich, but to carry out the liquidation here or to impose a regular sentence. Anything else is a burden of the Reich and continually leads to difficulties. Here we have an entirely different form of treatment and this form must be maintained." *Hans Frank's Diary* edited by Stanislaw Piotrowski, Panstwowe Wydawnictwo Naukowe, Warsaw, Poland, 1961, p. 61.

Chapter 5
The Orion

In the summer of 1958 I found myself in Santa Monica, where I had a job as a consultant for the RAND Corporation. This had followed a year at the Institute for Advanced Study in Princeton. In a way, RAND resembled the institute, a diverse group of very intelligent people in a setting of considerable physical beauty and comfort. There was one major difference: the RAND Corporation was at that time almost entirely devoted to strategies for winning the Cold War. The physics group, of which I was a part, was focused on nuclear weapons and their effects. One of its stars was the late Herman Kahn, who was trying to make us feel comfortable with the concept of "mega-deaths"—ours and the Russians. When, a few years later, I saw *Dr. Strangelove* in which the RAND Corporation had been renamed the BLAND Corporation, it rang a bell. This was not too surprising, since Stanley Kubrick had saturated himself in RAND reports and, indeed, Khan was one of the models for Strangelove. The accent came from the photographer Weegee (Arthur Fellig) who was on the set. In any event, life on the beach in Santa Monica aside, the whole atmosphere soon began to weigh heavily. Once a week there would be a packet from the Institute with mail and a gossipy letter from the secretary in the building that housed many of the physicists. In one of her letters she had some news about Freeman Dyson, whom I had gotten to know. She said that he had gone to a bullfight in Tijuana, where previously he had been bitten by a dog, and was working on an atomic bomb-powered spaceship at General Atomic—now General Atomics—in La Jolla, San Diego, south of Los Angeles. I wrote to Dyson summarizing what I had heard about him, and said that if any of these three things was true he was having a better time than I was. To my surprise, a few days later he called to invite me to come to La Jolla to join the spaceship program, which he said was called Project Orion. I agreed at once and piled my belongings into a newly acquired Hillman Minx convertible for the drive south. I did not have the foggiest idea what to expect.

As it happened, I got in on the ground floor. The project had begun officially that summer, and the company had just moved into a magnificent new complex near the Torrey Pines State Park. General Atomic had been founded in 1955 as a division of General Dynamics, one of whose specialties was building nuclear submarines. General Atomic—GA—which was to explore various aspects of nuclear energy, had been the brainchild of Frederic de Hoffman, a remarkable operator was a close collaborator of Edward Teller. Everyone referred to de Hoffman—sometimes to his face—as "Freddy." Freddy, who died in 1989, collected physics superstars for his enterprise the way others collect postage stamps. Dyson, who first came out in 1956, was an early recruit. Dyson has his name on a patent for what was, as far as I know, the first product to make GA any money: the TRIGA reactor. TRIGA stands for Training, Research, Isotopes, General Atomic. It was designed to be intrinsically safe, immune to the vagaries of human stupidity. The basic idea, which I think was Dyson's, was to arrange things so that as soon as anything happened to heat up the reactor core, the fission reactions immediately shut down. If you have a conventional reactor in which the fuel elements are surrounded by, say, a water moderator, which also acts as a coolant, then if the core heats up drastically, although the water stays relatively cool, the whole thing can run away. With the TRIGA one incorporated the moderator—say, hydrogen—within the fuel elements. When these are heated they will act to heat the fission neutrons, which will very rapidly be taken out of the energy range where they are effective in producing further fissions in a chain reaction. The chain reaction stops before any damage can be done. Fuel elements that are made of alloys of uranium hydride and zirconium hydride are the materials of choice. The GA website extols the virtues of the TRIGA, for whose invention they seem to credit Edward Teller. Teller did have the idea of designing a super-safe reactor, but the technical work was left to others. The first prototype went into operation a few months before I got to GA that summer, and I used to watch the blue Cerenkov radiation—radiation produced by electrons that are moving faster than the speed of light in a medium like water—coming from the water-filled swimming pool in which the operating fuel elements sat. By now, something like 70 have been sold. They are used in schools and hospitals all over the world.

As I said, I joined Orion not long after the project began. What I did not fully realize at the time were the actual steps that got it going. These—and much more—I learned from a book, *Project Orion*, by Freeman Dyson's son George, who was five when I first met him that summer. George is not a physicist, but he has studied with the best, including, of course, his father. He manages a level of objectivity that I think is admirable. There is a difference of opinion even now as to whether it ever made sense to take on the Orion project, and George—I will call him and Freeman by their first names to keep them apart—states the pros and cons fairly. Speaking for myself, I can tell you that when I first learned what was being proposed I thought it was crazy. Before I explain my reaction, which you may well come to share, let me briefly describe what was involved. We will fill in some of the details later.

In 1946, the Polish-born mathematician Stanislaw Ulam, who later collaborated with Teller to design the first successful hydrogen bomb, began thinking about how one could harness the energy in atomic explosions to propel a large object into space. What use he expected to make of such an object is not clear. Ulam, whom I used to refer to as the "mother-in-law" of the hydrogen bomb—he died in 1984—had a Mephistophelian side, so it probably had some sinister application. In any event, in 1955 he and another Los Alamos scientist named Cornelius Everett wrote a classified Los Alamos document called *On a Method of Propulsion of Projectiles by Means of External Nuclear Explosions*. As the title makes clear, they were thinking of propelling projectiles, and not people. The idea was to drop atomic bombs from the bottom of the vehicle, which would be set to explode at say 50 meters below it in some timed sequence. Accompanying these bombs would be disks of a lightweight plastic, which would be ejected separately above the explosion. When the explosion occurred it would transfer its momentum to the disk, which would be vaporized, and this vaporized material would in turn transfer its momentum to the bottom of the projectile, propelling it along in a series of bursts. A saucer-shaped vehicle was envisioned with a diameter of some 10 meters and weighing perhaps 12 tons. It is unlikely that anything would have come of this chimera if two things had not happened. On the one hand, in October of 1957, the Russians launched the first of two *Sputniks*, followed by another a month later carrying a dog named Laika. The second thing, which was crucial, was that a Los Alamos physicist named Theodore B. Taylor ("Ted"), who was a nuclear weapons designer by trade, had an epiphany. In this vision the unmanned vehicle of Ulam and Everett became an interplanetary ocean linear—like the Queen Mary—which Taylor called the Orion because it sounded somehow astronomical. (George dedicates his book to Ted Taylor, who died in 2004.)

Transporting people immediately raised an entirely new set of problems, which Ulam and Everett had not had to deal with since their hypothetical vehicle was unmanned. In brief, how do you keep the passengers alive while atomic bombs are exploding in their "basement" every few seconds? Indeed, why does the ship not simply vaporize in its entirety? These were just the sorts of questions that had persuaded me upon first hearing of the Orion that it was literally lunatic. I needed to be educated. My two teachers eventually became Ted Taylor and Freeman Dyson, from whom I also learned something of the history of the project.

Soon after de Hoffman founded General Atomic he set out to recruit Los Alamos scientists, and Taylor was one of his catches. In 1957 the two got a small contract from the Atomic Energy Commission, which enabled them to get started, and, more importantly, to have access to highly classified material on nuclear weapons. To work on the Orion you needed what was known as a Q clearance, which required a pretty rigorous inquiry by the FBI. As it happened, I had gotten one when I spent the summer of 1957 working as an intern at Los Alamos. This experience, as I will now explain, made me even more skeptical of the Orion. I hasten to add that I was not assigned to work on nuclear weapons at Los Alamos. In fact, I ended up working on something having to do with the decay of mesons, a problem in elementary particle physics for which no clearance was necessary, as it had no conceivable applications to anything. But I did get to watch two aboveground nuclear explosions in the Nevada desert. During the summer I had learned that if I paid my own airfare from Los Alamos to Las Vegas I would be allowed to visit the test site at Mercury some 65 miles from the city and see some tests. I agreed to this, and on August 30 in the company of a senior colleague, the late Francis Low, and the head of the Theory Division at Los Alamos, Carson Mark, flew to Nevada.

After the obligatory night playing blackjack in a local casino, I witnessed at dawn the first of these explosions. I have never been able to forget the impression this made. Later that day Carson Mark took Low and myself on a tour of the site.

We drove over some parts of the desert that were still radioactive and on which the sand had been fused into a sort of green glass. We climbed a 500-foot tower at the top of which was the bomb that was being assembled for the next day's test. It was a very large object, but how much was bomb, and how much superstructure, I could not tell. Then we were taken to a concrete blockhouse which, from the outside, seemed pretty undistinguished. But upon opening the door we found ourselves in a storage room that held enough nuclear weapons to destroy a country. The "pits" — the central spheres of plutonium that constitute the explosive material—were lined up on shelves waiting to have the covering of high explosive that triggered the implosion of the plutonium sphere glued on them. A man was in the process of doing this on a bomb while his wife or girl friend sat beside him knitting. To add to the surreal quality of all of this, Carson handed me a pit, which he took off the shelf. It was about the size and weight of a bowling ball and was warm to the touch. You can imagine my feelings, especially when Carson said it would be just as well not to drop it.

Having seen the explosion of nuclear weapons first hand, the idea of dropping perhaps hundreds of them, each weighing with its peripherals a few hundred pounds, from the back of a space vehicle seemed to me to border on madness. But a lot of things, including the internal combustion engine, until they are actually constructed, seem to border on madness. The internal combustion engine produces brief heat pulses at temperatures that would melt the material of the engine. They don't because the pulse is too brief. As it happened, one of Ted Taylor's specialties was designing compact and highly efficient bombs. The details of how this is done are still classified, but in his book, George tells us that the pits of these bombs are about the size of baseballs. They also have smaller yields than conventional bombs. The bombs I saw go off in the desert had yields equivalent to tens of thousands of tons of TNT—the sort of bombs that destroyed Hiroshima and Nagasaki. The "classical" Orion was to use bombs that ranged from a half a kiloton down to perhaps 50 tons, depending on where in the mission one was. Takeoff from Earth, for example, made use of ambient air as the propellant, and called for smaller yield bombs. I should point out that the object of this Orion was the exploration of plan-ets such as Mars. At one point Freeman told me that if it was done right, the cost of transporting cargo to Mars would be less than the going rate to transport cargo by air to Europe. Actually, Freeman had his eye on Enceladus, one of the moons of Saturn, which has an escape velocity of less than 400 miles an hour and probably contains water in the form of ice.

Back to the bombs: In a typical nuclear explosion of the kind I witnessed, the explosive force is pretty much uniform in all directions. But Taylor and others had developed directional nuclear explosives in which the bomb debris—the radiation and the rest—would come out in a well-directed cone. This could be focused on the propellant, which in this version of the Orion was part of the bomb package itself.

But that still left the next step. The propellant was to be driven against a circular metal plate with a diameter of perhaps a hundred feet. This plate was called the "pusher." Since the crew, living in relative comfort, were above the plate, they would have to be insulated both from the radiation and from the shock of being accelerated every second or so at about a thousand times the acceleration of gravity g, 32 feet per second per second. The pusher was designed to shield the crew from radiation, whereas smoothing out the ride, to perhaps an average acceleration of 2g, was to be achieved by a system of shock absorbers. It was remarkable to see Freeman, one of the most mathematically accomplished of all theoretical physicists, rolling up his sleeves and designing shock absorbers. How they would have worked in practice I do not know. But the really difficult problem was ablation: How much material would each explosion remove from the pusher? These explosions produce radiation with temperatures vastly in excess of the melting point of any material from which a pusher can be made. But, as Freeman pointed out to me, this is also true of the internal combustion engine, at least for a short time during part of its cycle. This is the key. You can stand these temperatures if they last very briefly. *How* briefly—the number of microseconds—depends in the case of the Orion on the opacity of the material from which the pusher is made.

In the course of the seven years that the Orion project lasted, I think it is fair to say that every theorist who worked on it took a crack at the opacity and ablation problem. One of the first was the late Marshall Rosenbluth, the very distinguished theoretical physicist, who noted that we were dealing with an extremely odd regime: hotter than the surface of the Sun but cooler than a bomb—a regime whose opacities neither astronomers nor bomb physicists had explored. When I got to La Jolla, Freeman was taking his own crack at the opacity problem. As usual, he had found an entirely novel way of looking at it. By making an ingenious use of the commutation relations between momenta and coordinates, and summing over all the spectral lines, he had found a sum rule that led to an upper bound for the total opacity. As far as I know, no one had ever thought to try this. In fact in 2003 we finally published it. (Publication of the Astronomical Society of the Pacific, Vol. 115 (2003) 1383–1387) My job was to take the numerical dipole matrix elements from tables and to add them up to see how rapidly the sum rule was saturated.

Various recollections come back. The first has to do with a lovely Southern California weekend during which, instead of playing tennis, I was in my office adding up matrix elements on a Marchant calculator, an electromechanical calculator that has now gone the way of the dinosaur. Freeman walked in and noticed that I was rounding off the four digit numbers on the table to three. He delivered a little homily about how errors could accumulate, and recalled that he had once done a calculation dropping what he thought were small errors along the way only to end up with an answer that was a hundred percent wrong! I redid my work with four digits. Remarkably, the sum rule seemed to work. I went into Freeman's office to announce this and he said, somewhat superciliously I thought, "Yes, it is remarkable that [q,p] = ih", this being a fundamental relationship between position and momentum in the quantum theory and which was the basis for the sum rule. At this point his chair fell over, and he was momentarily trapped underneath it. I told him I thought it served him right for having an attitude. When he had recovered he told me that he had recently discovered that the computer codes that were being used at RAND and the weapons laboratories appeared to produce results that violated the sum rule upper bound. This seemed fairly serious, so I suggested that we go north

to Los Angeles, where I would arrange with my former colleagues at RAND for
Freeman to give them a lecture on the subject. Soon afterwards this came to pass,
but I do not think the lecture was a great success. There were two reasons for this.
RAND had a mission, which was to plan for nuclear war in the age of missiles that
was now upon us. The Orion, with people flying around the Solar System, must
have seemed like a project that lacked gravitas. But there was another reason—and
this haunts the pages of George's book: secrecy. Freeman was confronted with a
large audience whose security clearances he had no way of knowing. Since he was
scrupulous about such matters, the lecture was necessarily somewhat incoherent.
I might note that a little earlier Feynman had been invited to work on the Orion
project and had refused. He said that he did not want to work on anything that was
classified, since if he got excited about it he would naturally want to talk to everyone. George ran into the same problem. Many of the people he interviewed could
only hint at the details.

It is not clear how much sense this makes now. Certainly one does not want to
give terrorists information that would help them to make small efficient bombs, but
my guess is that most of the Orion file could be released safely. George includes an
11-page list of documents produced by the Orion group. It is, he says, incomplete,
because of classification restrictions. He cannot even give the titles. The three
reports I wrote—two of them with Dyson—on opacities are there, including a long
summary that we presented in July of 1959: *The Continuous Opacity and Equations
of State of Light Elements at Low Density*. It was unclassified because someone had
decided that calculations of the opacities of elements lighter than iron were in the
public domain. By this time there were full-time experts on opacity at General
Atomic who were calculating in great detail what would happen. I think that some
of this was published in the open literature, but Freeman never published his sum
rule. It took, as I said, until 2003 before we published our joint work. As things
evolved, it became clear that the ablation situation could be vastly improved if one
coated the pusher with oil just before each explosion. It was also understood that
there were other effects that were exceedingly difficult to calculate—impossible,
with the computers that were then in hand. George notes that long after the Orion
died some of the computer codes that had been developed were resurrected and
found to be very useful.

By the middle of 1958, General Atomic had been awarded a contract for $999,750
by the Advanced Research Projects Agency—ARPA—a defense department entity
that was supposed to coordinate work on far-out ideas that might, or might not, play
a role in national defense. NASA did not yet exist, and, even if it had, it is unlikely
that it would have involved itself in something that required nuclear weapons. One
thing the contract enabled the project to do was to hire consultants like myself.
I recently came across my agreement with GA. It stipulated that I was to receive $50
a day, which, I learned from George's book, was the bottom of the scale. At the time,
it seemed to me like a fortune. My annual stipend from the Institute for Advanced
Study was less than $5,000. But, more important, the contract enabled the project to
begin an experimental program. Ultimately the idea was to test things like the ablation calculations with actual nuclear weapons. For reasons we will explain, this

never happened. What did happen was the construction of model Orions designed to fly, using a small number of chemical explosions. This also started in the summer of 1958. It reached its apogee, in all senses of the word, when a model with a 3-foot "pusher" attained an altitude of some 185 feet above Point Loma near La Jolla, from where it had been launched. It returned safely to Earth thanks to a parachute. Neither Freeman nor I was there to see it, but a film was made. In watching it, it somewhat strains credulity that this "put-put" was meant to chronicle the opening of a new Space Age, but I imagine films of the early chemical rockets would have made the same impression. The next step in the Orion would have been with atomic bombs and larger pushers, but it never happened.

There are several reasons for this. George discusses them in his book. In the first place, no one could fit the construction of a 4,000-ton vehicle, powered by hundreds of atomic bombs, into an established government entity. It was not exactly a weapons system and it was not exactly the peaceful exploration of outer space, although it was a bit of both. It became an administrative orphan. But long before this happened, several people on the project, including Freeman, became disillusioned. In Freeman's case this had to do with fallout. It can have escaped no one's attention that, if this vehicle had been launched from Earth, it would have produced a good deal of radioactive fallout. Freeman did some calculations and came to the conclusion that a Mars mission would induce at least one death from the effects of this radiation on innocent bystanders. I naively thought that this could be avoided if the Orion was first boosted out of Earth's atmosphere by chemical rockets, but Freeman noted that as long as one was in range of Earth's magnetic field it would trap the charged fission fragments, which would eventually fall back to Earth. The only way to beat this was to launch it from one of Earth's magnetic poles, adding to the general complexity of an already very complicated problem. On top of all this, it was a time when people's disgust with things nuclear was beginning to outweigh arguments about national security. The test ban treaties were being negotiated, and no one was about to make an exception for the Orion. Things like the pusher ablation could have been tested underground, but no money was forthcoming to do this. Finally, in 1965, the project died when the funds dried up. One of the more distasteful aspects of its death throes, as George chronicles in his book, was the attempt by people to invent military applications for a project whose raison d' être had always been the peaceful exploration of space. By this time both Ted Taylor and Freeman had long since moved on to other things. During its seven years, the Orion project had cost some $10 million. George notes that this was about the same as Stanley Kubrick's film *2001: A Space Odyssey*, which appeared the same year. Kubrick decided to use the name "Orion" for his space shuttle, but his Orion did not use nuclear weapons.

Is there a future for the Orion? This depends on how you feel about manned space travel. If you are an enthusiast, something like the Orion is probably the only way to go in the very long run. You can see this from what is called the specific impulse, which is a measure of the time it takes a particular quantity of fuel to generate a specific amount of thrust. More exactly it is defined to be the exhaust velocity of the propellant divided by the gravitational acceleration. You want to tune this to the needs of the mission, so that mass and energy are economized. This can be done on the Orion by adjusting the yield of the bombs and the mass of the propellant. As we know, you can get unmanned vehicles to Mars directly from Earth with chemical rockets, but not with people. This seems finally to have dawned on NASA. In his book George reports that they are once again looking into Orion-type vehicles that might be assembled in space. It would be a fine way of disposing of unwanted nuclear material. George assumed that they would have collected our old GA reports to study, but apparently not. George found himself in the odd position of sending NASA 1,759 pages of copies of our old reports by two-day UPS. There is something quaint about this: UPS and Orion, the human condition.

Freeman spent over a year in La Jolla, and was for a time considering moving there permanently. On one of my visits he told me a remarkable story. Not long before my visit, he been arrested for "walking" in La Jolla. He had gone out for a Sunday stroll and was stopped by a policeman. Freeman added off-handedly that, prior to his walk, he had broken his glasses, so he was wearing a pair of motorcycle goggles. When asked for identification by the officer he produced a red and black badge from Los Alamos, with his picture and fingerprints, stating that its bearer was entitled to receive top secret defense information. He was released at once. One can only wonder what went through the officer's mind. And one can wonder what the officer would have said if Freeman had gone on to explain that he was in La Jolla to design a spaceship that was propelled by atomic bombs.

Chapter 6
Tales from South Africa[42]

Then Indonesia claimed that they Were gonna get one any day.
South Africa wants two, that's right: One for the black and one
for the white! Who's next?

—Tom Lehrer, "Proliferation," 1965

Tom Lehrer wrote this song in 1965. Four years later the so-called "Atomic Energy Board" of South Africa began the exploration of the manufacture of nuclear weapons for what they claimed to be for peaceful purposes. They used the rhetoric of our so-called "Plowshares" program—an inspiration of Edward Teller designed basically to test nuclear weapons under the guise of peaceful applications. The South Africans chose to go the route of uranium rather than plutonium. The country has a good deal of natural uranium. The uranium the South Africans used was a byproduct of gold mining. They modeled their design after our Hiroshima bomb, Little Boy, although they substantially improved on it. This was what is known as a "gun assembly" device. A subcritical mass of uranium is formed into a projectile, which is fired by high explosives into a subcritical mass target. The two are then conjoined into a super-critical configuration, which undergoes an explosive fission chain reaction. This design was considered so straightforward at Los Alamos that it was never fully tested before the weapon was dropped on Hiroshima. To work on the weapons themselves one had to have been born in South Africa, and to work on anything related to the weapons one needed a security clearance that excluded blacks.

Uranium from a mine is over 99 percent uranium-238. This isotope cannot be used in a nuclear weapon, since a suitable chain reaction cannot be developed. It is the isotope uranium-235 that is "fissile" and must be separated from the uranium-238 matrix. There are a number of ways that this "enrichment" has been done, but the South Africans used a method that no other country has, as far as I know, ever used. Like nearly all the methods it uses a uranium gas-uranium hexafluoride, each molecule of which has one uranium atom attached to six fluorine atoms.

A mixture of uranium hexafluoride and hydrogen is injected into something that is called a "Helikon tube." The gas, which is mostly hydrogen, is injected at

---

[42]I am grateful to Carey Sublette, Waldo Stumpf, and Peter Zimmerman for comments.

J. Bernstein, *Physicists on Wall Street and Other Essays on Science Society*,
© Springer Science + Business Media, LLC 2008

supersonic speeds. The curves of the tube spin the gas as if it were in a centrifuge. The gas is acted on by a centrifugal force due to the spin. The force on the heavier isotope is greater and that portion of the gas continues on, while the light isotope, uranium-235, is siphoned off. The method is sometimes called a "stationary centrifuge" since it uses the same centrifugal forces as would be used in an actual centrifuge. In practice (compared to centrifuges) it is very energy consuming and inefficient. Despite some serious problems by 1982, the South Africans had produced enough highly enriched uranium to make six Hiroshima-like nuclear weapons. Some years earlier the then defense minister P. W. Botha had made it clear that these weapons were not for peaceful applications but for deterrence.

It is an interesting, and not an entirely settled, question, whether any of these weapons were tested. Certainly components of the gun assembly were tested using unenriched uranium to see if they worked. They did. But was highly enriched uranium ever used?

The South Africans deny it and deny they ever collaborated with the Israelis in such a test. But there is the matter of the Vela satellite. A fission weapon has a very characteristic signature. During the first microsecond the chain reaction produces all of the energy. The temperature of the gas—the left over uranium, etc.—is in the order of millions of degrees, like the interior of the Sun. A blinding flash of radiation is given off. But most of the energy is given off in the form of X-rays that are not visible. These plough into the surrounding atmosphere and knock the electrons off the atoms that make it up. This creates a charged "plasma," which blocks further radiation from the interior of the growing fireball until it has cooled to where the electrons can recombine with the atoms. That begins to happen after about ten thousandths of a second. There is then a second bright flash, which is less intense but of longer duration. These two flashes, separated by milliseconds, are absolutely characteristic of a nuclear explosion. A detector on the Vela satellite—the "bhang-meter"—had been designed to pick these flashes up and had, until September 22, 1979, an excellent record. On that date, one of the Velas picked up such a signal that seemed to come from the southern tip of Africa. But no other reliable collaborative evidence such as fallout has ever been recorded, and no one who was a witness to this test, if that is what it was, has ever come forward, either. It remains a mystery, although the scientific consensus was that it was a "zoo" event—a freak—and not a nuclear explosion.

Even more remarkable was the decision in September of 1989 by the then president of South Africa, F. W. de Klerk, to abandon the entire program. De Klerk made some pious remarks about South Africa wanting to join the world community. But he must have realized that apartheid was about over. Indeed, Nelson Mandela was released the following year. Tom Lehrer's "One for the black, and one for the white," was about to become "six for the African National Congress."

Dismantling a nuclear program is not so simple. It took nearly four years before everything from the weapons to the test sites to the enrichment program was gone. Under the supervision of the International Atomic Energy Agency most of the highly enriched uranium was melted down so that the form in which it had been used in the weapons could not be reconstructed. The exact amount that the South

Africans made has not been released, but it was enough for seven weapons—some 400 kilograms. Some of this is being used in the Safari reactor, which makes medical isotopes, and the rest is monitored by the International Atomic Energy Agency. The South Africans even shredded the weapons designs so that no one else could use them. Although the South African program was not large, as these things go, it did involve about a thousand people over the years. But even after 1993, when the official program was over, there were still offshoots. Around the year 2000 A. Q. Khan, the Pakistani nuclear arms dealer, sold a whole turn-key package for nuclear weapons to Libya. The problem was he couldn't deliver. He had to outsource. South Africa had a residue of experts left over from their program. In particular there was a company in Vanderbiljpark named Tradefin Engineering and run by a man named Johan Meyer. This company manufactured components for centrifuges. They began, under Khan's aegis, to sell them to a destination that turned out to be Libya. In 2004 Meyer was arrested and charged with trafficking in nuclear weapons. He said that he had no idea where the material his company was manufacturing was going. I was somehow reminded of other Tom Lehrer lyrics, these about Wernher von Braun: "'Once the rockets are up, who cares where they come down. That's not my department,' says Wernher von Braun."

Chapter 7
A Nuclear Supermarket

> *'Oxygen' is basic to life, and one does not debate its desirability; the nuclear "deterrence" has assumed that life-saving property for Pakistan.*
>
> —General Mirza Aslam Beg, Chief of Staff of the Pakistan Army from 1988 to 1991[43]

> *Dr. A. Q. Khan and his scientists have given the country a credible deterrent for a paltry sum of money. What they have in their accounts is what I call gold dust—they have not taken the government's money. If a scientist is given 10 million dollars to get the equipment, how would he do it? He will not carry the money in his bag. He will put the money in a foreign bank account in somebody's name. The money lies in the account for some time, and the mark-up that it fetches may probably have gone into his account. It is a fringe benefit.*
>
> —General M. A. Beg[44]

In the spring of 1969, I received word that my appointment as a Ford Foundation Visiting Professor of Physics at the University of Islamabad had been cleared by the government of Pakistan. I had actually gotten the appointment several months earlier, but there had been a not entirely peaceful change of governments in Pakistan. Ayub Khan was forced to resign and had been replaced by Yahya Khan. ("Khan" is a common Pathan name that was sometimes taken by non-Muslims who rebelled against the British.) Yahya had declared martial law and was cracking down on the universities, which is why my clearance had been delayed.

Now that I was actually going I gave some thought as to how I was going to get there. The obvious answer was to fly, but that seemed like a missed opportunity. I had read some accounts of people driving from Europe to India. There was even an extraordinary account by one Dervla Murphy who, in 1963, had ridden her bike from Ireland to India. In her book *Full Tilt*[45] she says that since she was traveling

---

[43] This is quoted in Shopping for Bombs, by Gordon Corera, Oxford Press, New York, 2006, p. 74.

[44] Corera, op. cit., p. 146.

[45] In 1987 it was re-issued as a paperback by Overlook TP.

J. Bernstein, *Physicists on Wall Street and Other Essays on Science and Society*, © Springer Science + Business Media, LLC 2008

alone, when she arrived in a town at night, her first action was to go to the local police station and ask to sleep there. I decided that I was going to drive.

This required several steps. First there was the choice of vehicle. That was fairly obvious. There existed a converted Land Rover called a Dormobile, which slept four and had a gas stove. It preserved all the rugged features of a Land Rover. The Ford Foundation people were somewhat surprised by my choice of transportation but agreed to contribute what they would have spent on an airplane ticket towards the purchase of the Land Rover. The rest of the money was loaned to me by the *New Yorker* magazine on the condition that after I sold the Land Rover I would pay it back. The fact that William Shawn, its editor, was willing to do this had to do with my choice of traveling companions. These were the Chamonix mountain guide Claude Jaccoux and his then-wife Michele. The three of us had spent some months together in Nepal two years earlier, and Shawn had liked the article that I had written about it. The Ford Foundation stipulation was that I was to appear in Islamabad by the first of October, when classes began at the university. Thus it was that the three of us left Chamonix on the fifth of September in the Land Rover, headed to Pakistan.

We went through the fairly recently completed Mont Blanc tunnel into Italy, drove to Venice, and then Trieste, where we stopped. I wanted to visit an old friend, Abdus Salam. Salam had been born in 1926, in a small town in Pakistan. He was a mathematics and physics prodigy, and after getting his undergraduate degree at the Government College at the University of the Punjab he won a scholarship to Cambridge, were he took his Ph.D. in 1951. His idea was to return to Pakistan to try to create a center for theoretical physics. He found that he simply could not get the support to do this, so he returned to Britain where, in 1957, he became a professor at Imperial College in London. In 1964 he helped to found what is now known as the Abdus Salam International Centre for Theoretical Physics in Trieste, whose purpose was to allow people from developing countries such as Pakistan to work on cutting-edge physics. For his own physics he shared the Nobel Prize in 1979 with Sheldon Glashow and Steven Weinberg. He was also a member of the Pakistan Atomic Energy Commission, which means he must have had detailed knowledge of Pakistan's nuclear weapons program. He died in 1996, and I never had a chance to ask him. He was pleased to see us and pleased that I was going to teach in Pakistan, but he found our method of getting there very odd indeed.

Our route continued through Yugoslavia, into Greece, then Turkey, past Mount Ararat, and into Iran. In eastern Iran we encountered several hundred miles of unpaved road and had our only flat tire. We had brought two sets of spare wheels. Then we entered Afghanistan and were stunned to find a veritable unused super-highway, which was the result of an American-Russian competition to build roads. We spent a little time in Kabul, including a side trip to see the giant Buddhas in Bamiyan that were later destroyed by the Taliban. Then we went over the Khyber Pass into Pakistan. I knew a good deal about the history of the Khyber Pass and was surprised that crossing it in an automobile was not very difficult. Once in Pakistan we headed for Rawalpindi, which is the old city that is a twin to the then new Pakistan capital, Islamabad. It was September 30. I made contact with my Pakistani hosts, who were very apologetic. Because of the student unrest of the previous spring the university was going to be closed down for a month. They had tried to notify me but, of course, I was en route. This was, in fact, wonderful news. We spent a month exploring the Northwest Frontier from Skardu to Chitral, even doing some treks close to the Afghan border as well as a side-trip to Swat. These must now be the infiltration routes of the Taliban. In the first week in November my abbreviated course began.

The teaching was very light so that I was free to do a good deal of traveling around the country. My hosts arranged a visit to the KANUPP nuclear reactor on the Arabian seacoast near Karachi. The reactor was then under construction. To get some insight into the Pakistan nuclear program, and to these programs in general, it is useful to present a little primer as to how a reactor works. Schematically there are four components. There are the fuel rods, largely composed of uranium in which the fission reactions take place. There is a coolant, which cools off the fuel rods to keep them from melting down. There is a heat exchanger, which takes the heat generated by the fissions and converts it into steam, which is used to power the turbines that generate electricity. Metaphorically speaking, a reactor is like a super-annuated kettle in which fission energy is converted into the energy of steam. Finally, there is something called a "moderator," which requires further explanation that I will now give.

For purposes of discussing fission, a heavy nucleus like that of uranium can be thought of as a liquid drop. The collective behavior of the dozens of neutrons and protons in such a nucleus resembles to a degree the collective behavior of the molecules in a liquid drop. When such a nucleus is bombarded by a neutron the "drop" is agitated, and if the energetic conditions are right it splits into "droplets." The larger droplets are nuclei of elements lighter than uranium. In the first fission reaction that was observed in December of 1938 by the German radio chemists Otto Hahn and Fritz Strassmann, the "droplets" were nuclei of barium and krypton. By atomic standards these fission fragments are produced with a high kinetic energy. It is this kinetic energy that is converted into heat. These fission fragments shed neutrons and, if more than two are produced in this process, then they in turn will generate new fissions, which produces a chain reaction. As I mentioned, the fission is started by a neutron incident on the "drop." What was unexpected, until Enrico Fermi discovered it in 1934, was that the slower these neutrons are the more readily do they produce nuclear reactions such as fission. This is a quantum mechanical effect. Neutrons do not behave like baseballs thrown at a window. The role of the moderator is to slow the neutrons. This happens because the neutrons collide with the nuclei of the moderator and give up some of their momentum in the collision.

The ideal moderator would be the nucleus of hydrogen—the proton. It and the neutron have about the same mass, so the proton can recoil like a struck billiard ball and take away a large fraction of the neutron's momentum. Hence ordinary water, whose molecule consists of two hydrogen atoms and an oxygen atom, would seem like the most desirable moderator. Indeed, the water could both cool the fuel rods and be transformed into steam.

However, there is a catch. There is always a catch. Some of the neutrons will combine with protons to produce the nuclei of "heavy hydrogen," which consists of two neutrons and a proton. These neutrons are then no longer available to the chain reaction. To compensate, in the so-called "light water" reactors the uranium in the fuel rods is "enriched." Uranium that comes from a mine consists predominantly of two isotopes, uranium-238 and uranium-235. Over 99 percent of natural uranium is uranium-238, whose nucleus has 92 protons and 146 neutrons. Uranium-235 nuclei have three fewer neutrons. Uranium-235 is much more readily fissioned than uranium-238, so to compensate for the neutron absorption problem with the light water moderator, in such a reactor the uranium is enriched so that its uranium-235 content is about 4 percent. This means that light water reactor technology inevitably gets mixed up with uranium enrichment and the possibility of enriching the uranium so that it can be used in nuclear weapons, which requires an enrichment of something like 90 percent.

There is another solution to the moderator problem, which is to use "heavy water," in which ordinary hydrogen is replaced by heavy hydrogen. The nucleus of heavy hydrogen—the "deuteron"—is not that much heavier than the proton, so it is a very effective moderator. In fact it is so effective that un-enriched uranium can be used as the fuel. Thus the problem of uranium enrichment is replaced by the much simpler problem of extracting heavy water from the natural water that occurs in the ocean or a lake.

The first people to make such a reactor were the Canadians. They began work on designing what is known as the CANDU-CANadian-Deuterium-Uranium-power reactor in the late 1950s. The first of these reactors to generate electricity went into service in Rolphton, Ontario, in 1962. When the Pakistanis wanted to build their first power reactor—the KANUPP—they hired the Canadians, who supplied everything including the heavy water. The KANUPP went into service in 1972 and ran without serious incident for its 30-year estimated lifespan. In 2002 it was shut down so that it could be modernized to extend its lifetime. During the entire era of its operation it was an open site subject to inspections. So far as I know, no plutonium—an inevitable consequence of any reactor operation—was ever re-processed. In short, the KANUPP reactor seemed to be a model of how such a facility should be operated.

On the other end of the spectrum is the reactor the Pakistanis built with the aid of the Chinese in Khusab in the Punjab. It, too, is a CANDU-type reactor, and apparently the Chinese supplied the heavy water for the moderator. The reactor went into operation in April of 1998. From the beginning all of its activities have been secret. From the evidence that has been collected, it is clear that the principal purpose of this reactor has been to produce plutonium. It apparently has also been used for the production of super-heavy hydrogen-tritium, whose atomic nucleus has two neutrons and a proton. This is significant because tritium can be used to make either hydrogen bombs or so-called "boosted" atomic bombs. The idea is that under the conditions created by a primary nuclear explosion the tritium and deuterium nuclei can "fuse." This fusion produces a nucleus of helium and gives off energy. But the most important thing that it does is to produce a neutron. Hence, in such a weapon a burst of neutrons is produced that greatly enhances the fission chain reaction. The Pakistanis acquired a tritium purification plant from Germany in

1987. There are also plutonium re-processing plants, the first one of which at Kahuta, near Islamabad, went into operation in 1984. It is estimated that enough plutonium has been produced to construct between 16 and 24 bombs. The uncertainty reflects the secrecy of the program.

Since I taught in Pakistan those many years ago I have often wondered how many of my students involved themselves in these activities. The temptations must have been great. The openings in universities in Pakistan are surely very limited, and many Pakistani scientists have emigrated. The ones that remained must have found opportunities in this kind of government work. Even the scientists who later chose to settle in Pakistan studied abroad whenever they could. When I was there, there was such a student about whom I heard nothing them. I doubt that very many people had heard anything about him. He was a metallurgist named Abdul Qadeer Khan. He was born in Bhopal, in what was then British India, in 1936. He emigrated with his family to Pakistan in 1952.

I have not been able to find out very much about Khan's early life. His father was headmaster of a school in Karachi. Khan did his undergraduate work in Karachi. I have not been able to learn how he was able to continue his education in West Berlin in 1961. Who sponsored him? But after a time in Germany, he moved to Holland, where he got a degree in metallurgical engineering from the Technical University in Delft in 1967. This was followed by a move to Belgium, where he took his Ph.D., again in metallurgical engineering, from the Catholic University of Leuven in 1971. Like other young Pakistani degree holders he decided to work in Europe. His thesis supervisor, Professor M. J. Brabers, heard about a job opening at the Physical Dynamics Research Laboratory (FDO) in Almelo, a small town in Holland. Khan moved there in May of 1972. As it happened at this very time the FDO, which was a subcontractor of Ultra Centrifuge Netherland (UCN), was getting into work on the most advanced centrifuges in the world. UCN was the Dutch subsidiary of a European consortium called the Uranium Enrichment Company (URENCO). The German branch had just come up with an advance design of a centrifuge. We will explain how URENCO came into being and why it had such a design later. The people at UCN needed someone who understood the material to translate the German documents into Dutch. Khan, who knew both languages, was perfect. It gave him access to the most classified material there was on the centrifuges that could be used to enrich uranium. What no one knew at the time was that this personable young Pakistani, who had married a Dutch-South African woman, had begun in 1974 to steal this material in order to bring it to Pakistan, where it became an essential component of the Pakistani atomic bomb program. In December of 1975, Khan made a hurried departure to Pakistan with blueprints for the centrifuges, which he turned over to the Pakistani Atomic Energy Commission after he went to work for them shortly thereafter. From that time until he was finally unmasked in 2001, Khan created a network that sold or exchanged information and nuclear weapons technology with countries such as Libya, Iran, and North Korea. Many of his operations were run out of Dubai, where interested customers could obtain a price list of the various offerings. The complete package, which included details as to how to make a bomb from enriched uranium and plutonium cost hundreds of millions of dollars. It was bought by the Libyans. The Libyans may have been short-changed, because the Khan network apparently sold them a somewhat antiquated design for a bomb.

These matters are described in detail in a book, *Shopping for Bombs*, by a Security Correspondent for the BBC named Gordon Corera.[46] It is a very interesting but ultimately depressing book, depressing because of the continuing havoc that A. Q. Kahn and his network produced in the attempt to control nuclear proliferation. Corera states at the beginning that he is not writing a biography of Khan. I wish he had included more biographical material, especially about Khan's early life. Apart from Khan's being a superlative hustler, I was never able to evaluate his abilities as an engineer. How much of the bomb technology did he actually understand? Before we try to answer that, let's consider something that Corera really does not deal with adequately, how this rather obscure Dutch company got itself into the business of manufacturing the most advanced centrifuges in the world.

The first person to use centrifuges to separate isotopes was the American physicist Jesse W. Beams of the University of Virginia. In the late 1930s he was able to partially separate the isotopes of chlorine. In 1941, with colleagues, he made the first successful separation of the uranium isotopes. This method was ultimately not used in the Manhattan project because it seemed more subject to problems than the other methods that were used. The next great leap forward in centrifuge technology came from an unlikely source—a detention center for German scientists captured by the Russians called "Institute A." One of the detainees was an Austrian physicist named Gernot Zippe. Zippe had served in the Luftwaffe during the war. He was captured by the Russians and first sent to a detention camp in Krasnogorsk. He was then transferred to Institute A, a research facility run by German detainees in Sukhumi, on the Black Sea in Western Georgia. The assignment of these detainees, many of whom had worked on the German nuclear energy program, was to provide a method, or methods, for separating uranium isotopes. Zippe joined a group working on centrifuge separation, that then proceeded to create the modern centrifuge. They produced almost frictionless bearings on which the centrifuge rested so that it used less power than a dim lightbulb.

Zippe remained in Russia until 1956, when he returned to Germany to join the Degussa Company. Degussa had an unsavory wartime history. It provided uranium metal for the unsuccessful German attempt to develop a heavy water reactor. The most dangerous parts of producing these metals were done by the forced labor of concentration camp detainees, some of whom died from their efforts. After it began producing the Zippe centrifuges, Degussat sold some of them to Iraq. Zippe spent the years from 1958 to 1960 in the United States, working with Jesse Beams. In 1964, the Germans formed a state-owned company to develop this uranium separation technology on an industrial scale. In 1970, the company was privatized, and that year it signed an agreement known as the Treaty of Almelo with similar enterprises in England and Holland to create a joint company, URENCO. That is why Khan was employed by UCN to translate the plans for the Zippe centrifuge from German into Dutch.

---

[46] Corera, op. cit.

To explain what happened next in Khan's saga I have to say a bit about the history of East and West Pakistan. When India was split in 1947, two Pakistans were created on opposite sides of India; these were separated by a thousand miles. From the beginning there was trouble. East Pakistan had a larger population, but the capital of the country—Karachi—was in the west. The people of East Pakistan, with good reason, felt economically exploited. There began a struggle for independence, which was held in check by the Pakistani army until 1971, when the East Pakistanis declared their independence and the Pakistani army was defeated by the Indians. Thus Bangladesh was created. The army in Pakistan has always had a special place of honor. It is what keeps the country from disintegrating. The spectacle of Lieutentant-General "Tiger" Nazri close to tears as he stripped his epaulettes and handed over his revolver to his Indian counterparts was so humiliating that the Pakistanis have never forgotten it. In December of 1971, Yahya Khan resigned, and Zulifikar Ali Bhutto became president. He had been proclaiming the need for a Pakistani nuclear deterrence for several years, even if, as he said, people had to eat grass to achieve it. Now that he was in power this was one of his first priorities.

In January of 1972, Bhutto called a council of advisors to decide on the next step. Pakistanis from the diaspora were summoned to join their local counterparts. A colleague of mine, the one who had brokered my appointment in Pakistan, was asked to go. As far as I know he refused. At the council there were voices raised against the idea of a relatively poor third-world country spending fortunes of money on nuclear weapons when there were so many social needs. As is usually the case, they were overruled. All doubts vanished when India tested its first weapon in May of 1974. This is when Khan decided to act. He wrote a letter to Bhutto offering his services. Remarkably it got to the president, and Khan was invited to present his case. This led to Khan's precipitous departure from the Netherlands, taking with him the details of the Zippe centrifuge. Once Khan began to work with the official body—the Pakistan Atomic Energy Commission—he decided that that centrifuge was much too cumbersome and persuaded Bhutto to set him up in his own facility, which he chose to locate in Kahuta, near Islamabad. From that point on, until his final downfall, he was accountable to no one, although he did collaborate with the army. This autonomy enabled Khan to set up his network, which sold or bartered nuclear technology. In his book Corera goes through the activities of the network country by country. Following are a few of the highlights, beginning with China.

In 1945, the memo that contained a blueprint for the Los Alamos plutonium implosion bomb, which had been given to the Russians by Klaus Fuchs, was sent to Beria and became the basis of the first Russian atomic bomb. The Russians, in turn, during a brief period when their relations with the Chinese were exceptionally good, gave this blueprint to the Chinese. The Chinese in turn traded it to Khan for the centrifuge technology. It took until 1998, before the Pakistanis successfully tested their first bomb, which, if Khan is to be believed, was a boosted uranium implosion weapon. The North Koreans traded missiles for the centrifuge technology. Airplanes from the Pakistani air force made the deliveries and pickups showing that Khan had the cooperation of the military. The civilian government appeared to have been kept in the dark. The Iranians bought centrifuge technology and even some parts. When the IAEA inspected one of these parts they found that it contained traces of uranium that had been enriched to 40 percent. One explanation was that the Iranians had a secret military program. The more likely explanation was that the part had been used in the Pakistani program. It was the commerce with the Libyans that finally brought the network down—if it is down.

For many years Colonel Gadafi had mixed feelings about acquiring nuclear weapons for Libya. But in 1995 he made the decision to go ahead. His representatives contacted representatives of the Khan network. Khan himself came to Tripoli several times. Gadafi decided to buy the package. This included the centrifuges, uranium hexafluoride gas to put in them for enrichment, and plans to make a nuclear weapon using the enriched uranium. In 1997, the Khan network delivered 20 so-called P-1 centrifuges of the Zippe type, which enabled the Libyans to get started. But then two problems arose. The first was that the network did not have enough material on the shelf to supply the Libyans. They needed to manufacture some, so they set up disguised factories in places such as Malaysia. The second problem was much more serious, indeed fatal. By this time the network had been penetrated. There were moles that have never been identified. All of Khan's activities were known. One might naively think that this would have been enough to bring the network down. The problem was putting enough pressure on the Pakistani government to do this. When the Russians entered Afghanistan we needed the Pakistanis to allow passage of the Muhajadin into Afghanistan. When the Russians left we needed the Pakistanis to help us fight the Muhajadin that we had created. To break the impasse something dramatic was needed, and this was supplied by the German-owned ship the *BBC China*. This ship had been engaged to bring nuclear material from Malaysia to Libya in October of 2003. It was tracked from the time it left port until it stopped in Italy, where it was boarded and its cargo seized. This was the smoking gun that could not be ignored.

Gadafi had had second thoughts about his nuclear program, and the seizing of the *China* was the last straw. He decided that he would use giving up his nuclear booty as a chip to trade for recognition by the United States and assurances that there would be no attempt at a regime change. There then commenced a series of negotiations, which often seemed like something out of Monty Python. On December 11, as the plane with the CIA team was about the leave Tripoli, the Libyans gave them a stack of several envelopes. These, it turned out, contained the blueprints for the bomb that Khan had sold them. As far as I know these have never been released, so that one does not know exactly what bomb these plans were for. It may have been for the un-boosted uranium fission weapon. With this out in the open, General Pervez Musharraf, now president of Pakistan, could no longer maintain the fiction that Khan's activities were a mystery to the government.

On January 31 Khan was relieved of his position as special advisor to the prime minister. On February 1 Khan confessed to Musharraf what he had done, and three days later he gave a speech to the nation apologizing. But for many he was still a hero—the father of the Pakistani atomic bomb. He was never tried. He was put under a comfortable house arrest incommunicado in Rawalpindi, where he is at the present day. Around the world his network was being dismantled, a process that is still going on today. In the summer of 2006, to take an example, a German named Gotthard Lerch, who was accused of helping to arrange the Libyan deal, went on trial in Mannheim. He had to be extradited from Switzerland. This was contested by Lerch's lawyers, and the trial has been suspended indefinitely. Others connected to the network have claimed that they knew nothing of what it was actually trafficking in and as far as they were concerned the transactions were perfectly innocent. How much, if any, of the network is still operating is undetermined.

In his book Corera speculates about Khan's motives. In the beginning it seems that they were patriotic. He seems to have lived modestly. But he clearly had a tendency towards megalomania, and by the end it seems to have been a mixture of power and greed. He built a palatial mansion on the shores of Rawal Lake near Rawalpindi. In doing so he violated the local laws by spilling raw sewage into the lake. When the local authorities tried to bulldoze the house, Khan's bodyguards shot the bulldozer operator. This is presumably the house that Khan is living in today.

Prior to Khan, the German physicist Klaus Fuchs was the greatest source of proliferation by espionage. As mentioned earlier, he stole the complete blueprint of the bomb that was dropped on Nagazaki. Whatever one may think of him, greed was not his motive. In a fascinating book, *The Man Behind the Rosenbergs*,[47] Alexander Feklisov, who was the Soviet agent who ran the activities of the Rosenbergs in the United States and Fuchs in England, describes Fuchs's adamant refusal to take money. Only in his last meeting with Feklisov, on April 1, 1949 (Feklisov died recently at the age of 93), did Fuchs accept money. His older brother Gerhard was suffering from tuberculosis and was in a very expensive sanatorium in Davos. Fuchs took the money to help out his brother.

---

[47] *The Man Behind the Rosenbergs*, by Alexander Feklisov, Enigma Books, New York, 2001.

Chapter 8
Ottavio Baldi: The Life and Times of Sir Henry Wotton

> *My next and last example shall be that undervaluer of money,
> the late provost of Eton Colledge, Sir Henry Wotton (a man
> with whom I have often fish'd and convers'd) a man whose for-
> reign Imployments, in the service of this Nation, and whose
> experience, learning, wit and cheerfulness made his company
> to be esteemed one of the delights of mankind; this man, whose
> very approbation of angling were sufficient to convince any
> modest censurer of it, this man was also a most dear lover, and
> a frequent practiser of the art of angling; of which he would
> say; of which he would say, 'Twas an imployment for his idle
> time, which was not idely spent; for angling was after tedious
> Study, a rest to his mind, a (cheerer of his spirits, a diverter of
> sadnesse, a calmer of unquiet thoughts; a moderator of pas-
> sions, a procurer of contentedness; and that it begot habits of
> peace and patience in those that profess's and practis'd it.
> Indeed, my friend, you you will find angling to be like the ver-
> tue of Humility, which has a calmness of spirit, and a world of
> other blessings attending upon it.*

—Isaak Walton (The spelling and the italics are from the original.)

> *Blest silent groves, oh may you be
> For ever mirths best nursery:
> May pure contents
> For ever pitch their tents
> Upon these downs, these meads, these rocks, these
> mountains,
> And Peace still slumber by these purling fountains:
> Which we may every year
> Meet when we come fishing here.*

—Henry Wotton[48]

---

[48] Walton, op. cit., p. 259–260.

J. Bernstein, *Physicists on Wall Street and Other Essays on Science and Society,*
© Springer Science+Business Media, LLC 2008

For me! —if there be such a thing as I—
Fortune-if there be such a thing as she—
Finds that I bear so well her tryanny,
That she thinks nothing else so fit for me—
But though she part us, to hear my oft prayers
For your increase, God is as near me here:
And, to send you what I shall beg, his stairs
In length and ease are alike every where.

The last stanzas of a poem by John Donne on the occasion of Wotton's departure to Venice as the British ambassador.[49]

Sometime around 1624, Isaak Walton, who was an iron monger in London, moved to Fleet Street, not far from St. Dunstan's in the West, the church whose vicar was John Donne. Walton had a genius for friendship and soon became part of the circle that included Donne and Henry Wotton, who had known Donne ever since they were young teenagers at Oxford. Walton soon became a fishing companion of Wotton's, whose life of Donne was never completed. Walton used information obtained from Wotton who died in 1639. Walton also wrote an affectionate profile of Wotton, which he included in his collection, *Lives*. It is one of the very few biographies of Wotton so, despite its hagiography, is very valuable.

On its face, a more unlikely pair than Wotton and Walton (Walton was some 25 years Wotton's junior), is difficult to imagine. While Walton, who had little formal education, had a parentage of no distinction, Wotton, who was one of the most cultivated men of his age, came from a family of both antiquity and nobility. His father had refused a knighthood from Queen Elizabeth. Wotton and his three stepbrothers were knighted. Francis Bacon was related. When Walton met him he was in his late fifties and had recently been appointed provost of Eton. But he had already led several lives.

After Oxford he lived abroad for six years, perfecting his languages with an eye towards a diplomatic career. He made the mistake of befriending and becoming secretary to Robert Devereaux, the second Earl of Essex. He accompanied Essex on his expeditions to Cadiz and the Azores and on his ill-fated expedition to Ireland. This led to a plot against Elizabeth I that cost Essex his head in 1601. When this was unfolding, Wotton was back in England but managed to escape to the Continent to avoid being implicated. Actually, Wotton had nothing to do with the plot but felt that this fine distinction could easily get lost in the emotions of the moment. He might well have remained abroad indefinitely if he had not heard, while living in Florence in 1602, of another plot, this one to kill James VI of Scotland, who eventually succeeded Elizabeth to the British throne. The Medici grand duke of Tuscany Ferdinando I gave Wotton the mission of transmitting some letters to James warning him of the plot and instructing him how to employ the antidote to the poison that was to be used, in case he swallowed any. For the purpose of his mission, Wotton disguised himself as an Italian. He took the name Ottavio Baldi—revealing his real identity only to James.

---

[49] *The Lives, by Isaak Walton*, George P. Putnam, New York, 1848, p. 145

Needless to say this endeared him to the future king, who rewarded him with a knighthood and diplomatic missions, including an ambassadorship to Venice, which began in 1604 and lasted, with two interruptions, for nearly 20 years. His letters describing the Venice of that period have become a primary source for historians.[50]

One of the interruptions to Wotton's Venetian career began as an innocent jape that turned sour and might well have finished it off definitively. On his way to Venice in 1604, when he was taking up his post, Wotton stopped off to see friends in Germany. During an evening of merriment he was asked to write a sentence in his host's autograph book. Wotton wrote in Latin, "*Legatus est vir bonus peregrè missus ad mentiendum Reipublicae causâ.*" The English language translation of this, which is certainly what Wotton must have had in mind, with its delicious pun, may be one of the best-known aphorisms about the diplomatic service: "An Ambassador is an honest man, sent to lie abroad for the good of his country." Unfortunately there is no pun in Latin. "*Mentior*" means to tell a falsehood *tout court*. No one noticed this for almost eight years when a German writer named Jasper Scioppius somehow got a hold of the book in which Wotton had written his *bon mot*. Scioppius was a Catholic who despised King James and his religion. He wrote a scurrilous book, *Ecclesiastus*, and used Wotton's epigram—the Latin version—to demonstrate the low level of morality of King James's emissaries. When the king found out about this he was, to put it mildly, not amused. Wotton was recalled, and it took all of his skill, and no doubt reminders that he had once been Ottavio Baldi, to get him reinstated. He returned to Venice in 1616, and with another interruption that lasted from 1619 to 1621, he remained in Venice until 1624, after which he became the Provost of Eton and Walton's friend. It was something that occurred during this second interruption that first aroused my interest in Wotton.

To explain, we need to give a brief account of King James's marital history. In 1589, James married Princess Anne of Denmark in a ceremony performed in Copenhagen. Incidentally, he visited the observatory of the great Danish astronomer Tycho Brahe, his only foray into astronomy, which has some relevance to what follows. The marriage produced two children who survived past their teens: Charles, who succeeded his father and was eventually beheaded, and a daughter Elizabeth. In 1618, Elizabeth was married off to Frederic V, the Protestant elector Palatine of the Rhine. James saw himself as a Protestant island surrounded by a vast sea of hostile Catholics. In 1619, Frederic became the king of Bohemia. He was deposed the next year by a Catholic army of Spaniards and Bavarians. He and his wife were sent into exile in Germany, where he died in obscurity. James, both out of concern for his daughter and the fate of the Protestants in central Europe, tried to arrange some sort of negotiated peace between Catholics and Protestants prior to the deposition of his son-in-law. To this end he sent a mission to the Continent. Donne was assigned by the king to accompany the group as its spiritual advisor. The mission, which was a failure, wandered around Europe for close to a year.

---

[50] For a recent example see Gary Wills, *Venice; Lion City*, Simon Schuster, New York, 2001, where Wotton's letters from Venice are quoted extensively.

On October 23, 1619, it reached Linz. There then took place one of the oddest meetings it is possible to imagine, between John Donne and Johannes Kepler.[51] Our knowledge of this meeting is sketchy, but it must have been arranged by Donne. It is clear from Kepler's description of it—we know only of a single letter—that, apart from seeing Donne as a useful vehicle to convey his astronomical ideas to King James, he had no clue as to who Donne was, then or afterwards. This was especially ironic because in 1611, Donne had published in both Latin and English—anonymously—a biting anti-Jesuit satire that he called *Ignatius His Conclave*. In it he mentions Kepler explicitly. He even knew that Kepler succeeded Tycho in Prague after the latter's death. Kepler refers to this in print but had no idea who the author was, and, at their meeting, Donne did not explain. Kepler had wanted to give Donne a copy of his newly published—or nearly published—book *Harmony of the World*, with its remarkable mixture of perishable mysticism and enduring science. It was "nearly" published because the copies Kepler had available did not contain his dedication to King James. Kepler was a Protestant, and he saw in James a defender of the faith. There is no indication that James had any interest in Kepler's work. Enter Wotton.

After the failure of the mission of which Donne had been a part, the king turned to Wotton. In 1620, he appointed him a special emissary with the responsibility of looking after the interests of his daughter. Wotton seems to have had a real affection for Elizabeth. Indeed, he dedicated a poem to her. It begins:

You meaner beauties of the night,
That poorly satisfy our eyes
More by your number than your light,
You common people of the skies, —
What are you when the sun shall rise?[52]

In the summer of 1620, he, too, visited Kepler in Linz. But Wotton had a mission, on whose authority it is unclear, which was to invite Kepler to relocate in England. There are at least two extant letters in which Kepler speaks of this invitation and why he cannot accept it.[53] He describes his perilous situation with the country engulfed by the flames of war— "*Flagare vides incendium belli civilis in Germania*" —but he simply cannot bring himself to leave. He stayed in Linz until the city was actually burned down in 1626 and then wandered from place to place until he died in Regensburg, Germany, in 1630. But we owe to Wotton's visit one of the rare glimpses as to how Kepler must have actually appeared. It is contained in a letter that he wrote to his distant cousin Francis Bacon[54] from Vienna in December of

---

[51] The interested reader can find a more detailed account of this in my book *The Merely Personal*, Ivan Dee, Chicago, 2001.

[52] See Walton's *Lives*, op. cit., p. 153 for the full text of the poem.

[53] The Latin text of these letters can be found in *The Life and Letters of Sir Henry Wotton*, by Logan Pearsall Smith, Oxford Press, Oxford, 1907, Vol. II, p. 205. I shall have many occasions to refer to this work which I shall abbreviate as LPS.

[54] The relationship is a little complex. Bacon's mother, Anne Cooke, was the great-granddaughter of Sir Philip Cooke. He had married the daughter of Sir Henry Belknap whose sister was the wife of Wotton's great grandfather. LPS, Vol. II, has an invaluable appendix in which people connected to Wotton are identified.

1620. He seems to have promised to report on anything he had come across of scientific interest. Wotton had always been interested in science. He writes[55]:

"I lay a night at Lintz, the metropolis of the higher Austria, but then in a very low estate, having been newly taken by the Duke of Bavaria, who *blandiente fortuna* [by the blandishments of fortune], was gone on to late effects. There I found Keplar [sic], a man famous in the sciences, as your Lordship knows to whom I purpose to convey from hence one of your books, that he may see we have some of our own that can honour our King, as well as he hath done with his *Harmonica*. [Kepler's *Harmony of the World*, which he had dedicated to James.] In this man's study I was much taken with the draft of a landscape on a piece of paper, methought masterly done: whereof inquiring the author, he betrayed with a smile it was himself; adding he had done it *non tanquam pictor; sed tanquam mathematicus* [not as a painter but as a mathematician]. This set me on fire. At last he told me how. He hath a little black tent (of what stuff is not much importing) which he can suddenly set up where he will in a field, and it is convertible (like a windmill) to all quarters at pleasure, capable of not much more than one man, as I conceive, and perhaps at no great ease; exactly close and dark save at one hole about an inch and a half in the diameter, to which he applies a long perspective trunk, with a convex glass fitted to the said hole, and the concave taken out at the other end, which extendeth to about the middle of this erected tent, through which the visible radiations of all objects without are intermitted, falling upon a paper, which is accommodated to receive them; and so he traceth them with his pen in their natural appearance, turning his little tent around by degrees, till he hath designed the whole aspect of the field. This I have described to your Lordship, because I think there might be good use made of it for chorography [regional map-making]: otherwise to make landscapes by it were illiberal, though surely no painter can do them so precisely." This was Kepler's version of the *camera obscura*.[56]

It was this encounter, which I learned about from a colleague—Owen Gingerich, who is a renowned Kepler scholar—that aroused my interest in Wotton. I soon discovered that there is rather little written about him. As far as I have been able to learn, there are at most a handful of biographies, the most important of which is the one written by Walton, and the two volume work *The Life and Letters of Sir Henry Wotton*, which was published in 1907, by the Philadelphia-born English writer Logan Pearsall Smith. No one seems to have written a serious biographical study since. Considering how often Wotton's name pops up this, at first sight, is quite odd. I think that the explanation has to do with the difficulty in explaining what Wotton actually did. It is much easier to explain what he failed to do, which was to carry out most of the projects—at least the literary projects—that he began.

---

[55] LPS, Vol. II, p. 205. I have kept the original spelling in which, for example, "Linz" is spelled "Lintz" and "Kepler, " "Keplar."

[56] One would imagine that Wotton and Kepler must have spoken German, in which Wotton was fluent. Kepler and Donne must have spoken Latin, since there is no evidence that Donne spoke German and none that Kepler spoke English. It should also be noted that the artist David Hockney has generated something of a controversy by claiming that noted Renaissance painters used a *camera obscura* to make their drawings more precise.

A prime example was his biography of John Donne based on the relationship of which Walton wrote, "The friendship of these two I must not omit to mention, being such a friendship as was generously elemented; and as it was begun in their youth, and in an University, and there maintained by correspondent inclinations and studies, so it lasted till age and death forced a separation."[57] Since Walton knew both of them one must assume that he is right. Wotton, who outlived Donne by eight years, did gather material for a biography, which he never wrote.

Edmund Gosse, who was nothing if not industrious, commented on this in his two-volume study of Donne, *The Life and Letters of John Donne*,[58] which he published in 1899. He writes, "Wotton testified to this affection by undertaking to write Donne's life, and he applied to Isaak Walton to collect materials for him. 'I,' says Walton, 'did prepare them in a readiness to be augmented and rectified by his powerful pen.' (1640) In 1658 [long after both Donne and Wotton were dead] he added that he did 'most gladly undertake the employment, and continued it with great content, till I had made my collection ready.' We may easily reconstruct what happened in the matter. Sir Henry Wotton, a very magnificent amateur, with whose literary and piscatory recreations Walton was proud to be associated, remarked how welcome would be a life of their illustrious friend, the deceased Dean of St. Paul's; thereupon, Walton urged the Provost [Wotton was then the provost of Eton] to oblige the world with Donne's life 'exactly written.' Wotton, who shrank delicately from the printing press [not a condition that Gosse shared], would hesitate and yet admit the attractiveness of such a proposal; and would at last say that if Walton could obtain for him the material facts—their 'bodies' having been 'divided' —he would give the matter his best consideration. Walton thereupon set about gathering data together and setting down his recollections (many of them, doubtless, already submitted to paper), and Wotton was satisfied to find Walton so zealous. Of course when the Provost of Eton died 1639, no traces of his Life of Donne were forthcoming. Wotton was the most elegant of amateurs, but he was essentially an amateur."

Gosse goes on, "It has been customary to lament that Wotton was prevented from carrying out his design. On the contrary, we ought to rejoice that he did not, by the preparation of a vapid eulogy in the customary renaissance manner, interfere with the production of Isaak Walton's exquisite little masterpiece. There is nothing in the *Reliquiae* [In 1651, Walton published a collection of Wotton fragments that he had edited called *Reliquiae Wottonianae*], nothing in the sparse miscellaneous writings of Wotton to indicate that at any part of his life he possessed the unusual gifts required by a biographer [Gosse had written several], nor must it be forgotten that, when Donne died, Sir Henry, a wearied and asthmatic man, was already advanced in years. His irresolution in carrying out literary schemes was constitutional; he had proposed to himself a life of Martin Luther, a

---

[57] Wotton, *Lives*, op. cit., p. 136.

[58] *The Life and Letters of John Donne*, by Edmund Gosse, the 1959 reprint edition, Peter Smith, Gloucester. Pearsall Smith had access to some letters that Gosse was not aware of, but this does not change the fact that most of the correspondence seems to be lost.

complete manual of architecture, a history of the Reformation in Germany, and many other projects. He was the prince of those busy men of diplomacy, who are always hankering after a life of lettered repose, and who, when they secure it, do not know how to employ it."[59]

There is much truth in what Gosse says, but I think he misses the point. What makes Wotton worthy of our attention has to do precisely with the fact that he was the "prince of those busy men of diplomacy" and that he described his active life in a stream of wonderful letters; the letters that Smith collected systematically for the first time. In his delightful autobiography, *Unforgotten Years*,[60] which was published in 1937, when Smith was in his seventies, he tells us how he, an American from Pennsylvania, had set about to find these letters, most of which no one else had bothered to collect, or even read. Smith, who was born in Milville, New Jersey, in 1865, came from a long line of Quaker evangelists who had migrated here from England. His father had joined the glass business of his in-laws and prospered. The family moved to Pennsylvania. Smith describes his childhood and adolescence as a happy one, not unduly troubled by intellectual pursuits. However, he had an older sister, Mary, who was an intellectual and who took her brother in hand. In particular, she got him to read Walt Whitman who, as it happened, was living close by. Undaunted, Mary presented herself to Whitman in person, and he soon became a close friend of the family, often staying with the Smith's for weeks. Smith noted that on one occasion they came across an elegant Englishman who was looking for Whitman. They were able to show him where he lived and, indeed, Mary climbed up a ladder to knock on his window when he did not answer the door. The Englishman was, of course, Gosse, who became a good friend of Smith's. Mary went to Radcliffe, where she studied philosophy with the likes of William James. She then went to Oxford, where she married an Oxford don. After he died, she married the art historian Bernard Berenson. Smith's younger sister Alys married Bertrand Russell. Smith never married and neither did Wotton. When Wotton was in his fifties and about to return to England after his second tour of duty as ambassador to Venice, contemplating his much diminished purse and wondering what to do about it, he wrote to a friend: "And, peradventure, I may light upon a widow that will take pity of me."[61] Apparently, none was on offer.

Smith, too, went to Harvard with the understanding that he would go into the family glass business, which he did for about a year. He then decided that what he really wanted to do was to become a perpetual student. At first, his father was adamant about his remaining in the business as a debt to the family. But both his mother and older sister were equally adamant that Smith should do what he really wanted to do. The clinching argument was given by his mother, who said that, after all, the business belonged to *her* family and she was giving permission. This was so persuasive that not only did Smith senior give his son $25,000, on which he

[59] Gosse, op. cit., Vol. ii, p. 315.
[60] *The Unforgotten Years*, by Logan Pearsall Smith, Little Brown, New York, 1937.
[61] LPS, Vol. II, p. 130.

supported himself for several decades, but he decided to leave the business himself and move the rest of the family to England. Smith, now 23, enrolled in Balliol College in Oxford and then lived in Paris and elsewhere in Europe. It was there that he got the idea of writing about Oxford, which he eventually did: *The Youth of Parnassus and Other Stories of Oxford Life.*

Neither in his autobiography nor in his *Life and Letters* does Smith tell us how he first heard of Wotton, but in the *Life and Letters* he explains why he chose him as a subject, and in his autobiography he tells us how he went about his work. In the preface to the *Life and Letters*, he writes, "Among the contemporaries of Shakespeare an interesting but little-known figure is that of the poet and ambassador, Sir Henry Wotton. It is still remembered that he was the author of two or three beautiful lyrics which are to be found in every anthology; that he went as ambassador to Venice, and fell into temporary disfavour owing to a witty but indiscreet definition of his office; and that afterwards he became Provost of Eton, where he was visited by the young Milton, and where he fished with Isaak Walton, who quoted his sayings in the *Compleat Angler*; and wrote an exquisite portrait of his old friend. But behind the tranquil old age described by Walton lay many years of travel and participation in public affairs, much acquaintance with men, and with courts in foreign lands. The period indeed of Wotton's life covers the whole of what is known as the great age of Elizabethan literature, from the defeat of the Armada to the death of Shakespeare, and extends almost to the outbreak of the Civil Wars. It is hardly necessary, therefore, to apologize for the publication of his letters (which for the most part have remained hitherto unpublished), and a study, longer and more complete than any which has yet been attempted, of his career and character and public services."[62] What Smith could not have foreseen was that, in the century since he wrote his book, no one has written one that is more complete.[63]

In his autobiography he tells us how he was able to assemble the nearly 1,000 documents—letters and reports—from which he selected for his book. Incidentally, neither Walton nor his few successors had never really dated these documents correctly. Smith writes, "England has become the home of sport for many Americans, who come annually to this island for deer stalking, for fishing, and for the hunting of foxes. But there is another form of hunting which has occupied a good deal of my English leisure—the hunting, namely, for manuscripts of literary interest in English archives and old English country houses. I acquired my taste for this form of sport when I began to write the life of the old poet and ambassador and Provost of Eton, Sir Henry Wotton; and I spent some years in collecting his unpublished letters. The archives of the Record Office, the British Museum and the Bodeleian Library are easily accessible, and there are officials at these institutions ready and even eager to assist students in their labors."

---

[62] LPS, Vol. I, p. iii.

[63] For a recent example see the essay on Wotton in *Three Extraordinary Ambassadors*, by Harold Acton, Thames & Hudson, New York, 1983.

But this limited form of research did not satisfy Smith. He goes on, "When, however, I wished to pursue my hunting into the archives of private houses, I found that a much more elaborate method of procedure was required. It is quite useless, in my experience, to write out of the blue, so to speak, to great personages and ask permission to examine their muniments rooms, and inherited manuscripts. Either they will not reply, or they will send curt refusals. I think that they do not know themselves (not being literate people) what treasures they possess, or, if they do, they regard an unknown inquirer as a thief or gangster, with robbery as the object of his visit. I found it necessary, therefore, to procure some kind of personal introduction before writing to them. The plan I adopted was that of inquiring among the people I happened to meet if any of them knew, or knew anything about, the magnate whose manuscripts I wanted to examine; and once a personal relation of this kind was formed, however tenuous, all difficulties would at once vanish."[64]

The letters in Smith's collection begin in 1589, when the then 21 year old Wotton was preparing to leave England to live abroad. They end in 1639, the last year of his life. He concludes his final letter, which is to John Dynely, who had been his assistant secretary in Venice, by saying "Let us, howsoever, love one another, and God love us both."[65] Many of the letters are addressed directly or indirectly to the king. The former he often signs "Ottavio Baldi. " Unfortunately, none of the letters directly describes the episode which inspired Walton to disguise his name. We do know that in September of 1601, Wotton arrived in Scotland with a dispatch from the Grand Duke of Florence and a casket of antidotes to the poison which the king—then reigning as James VI of Scotland—was in danger of succumbing to. Fortunately, we do have Walton's savory account, which he must have heard from Wotton himself. He writes, "When Octavio Baldi [sic] came to the Presence chamber door, he was requested to lay aside his long rapier (which, Italian-like) he then wore and being entered the chamber, he found there with the King, three or four Scotch Lords standing distant in several corners of the chamber: At the sight of whom he made a stand; which the King observing. "bade him be bold, and deliver his message; for he would undertake for the secrecy of all that were present. " Then did Octavio Baldi deliver his letters and his message to the King in Italian; which when the King had graciously received after a little pause, Octavio Baldi steps to the table and whispers to the King in his own language that he was an Englishman, beseeching him for a more private conference with his Majesty, and that he might be concealed during his stay in the nation; which was promised and really performed by the King, during his abode there (which was about three months) all which time was spent with much pleasantness to the King, and with as much to Octavio Baldi himself, as that country could afford; from which he departed as true an Italian as he came hither."[66]

---

[64] *Unforgotten Years*, op. cit., pp. 239–242.

[65] LPS, Vol. II, p. 411.

[66] *Lives*, op. cit., p. 141.The spelling and punctuation are in the original.

Upon his return to Florence Wotton made a report in Italian of his impressions of the young king. He writes, *"E' di statura mediocre, di complessione gagliarda, le spalle larghe, il resto della persona in giu piu tosto sottile ch' altramente. Nelli occhi, et nella forma esteriore del viso, apparisce una certa bonta naturale al modesto."*[67] — "He is of average stature, a hardy complexion, broad shoulders, while the rest of his physique was slender rather than otherwise. In his eyes and in the external form of his face, he showed a manifest goodness and modesty." Walton describes the denouement to this episode. After the death of Queen Elizabeth in 1603, James came to England, where he was crowned James I. He discovered the presence of some of the people who had served the queen, among them Wotton's stepbrother Edward. The king asked Edward if he had ever heard of a Henry Wotton, only to be informed that it was his brother. Wotton was then summoned home from abroad, knighted, and made ambassador to Venice.

There are in Smith's collection 511 letters. Many of these are several pages long—dispatches really—and a number are in foreign languages such as Latin and Italian. They are meticulously annotated. It is a treasure trove in which the reader will have to find his or her way. Some readers may focus on Wotton's relationship to literature and the arts. For example, on July 2, 1613, he writes to Edmund Bacon, Francis Bacon's nephew and Wotton's closest friend, "Now to let matters of state sleep, I will entertain you at the present with what hath happened this week at the Bank's side. The King's players had a new play, called *All is true*, representing some principal pieces of the reign of Henry VIII, which was set forth with many extraordinary circumstances of pomp and majesty, even to the matting of the stage; the Knights of the Order with their Georges and garters, the Guards with their embroidered coat, and the like: sufficient in truth within a while to make greatness very familiar, if not ridiculous. Now, King Henry, making a masque at the Cardinal Woseley's house and certain chambers [a reference to some sort of cannon] being shot off at his entry, some of the paper, or other stuff, wherewith one of them was stopped, did light on the thatch, where being thought at first but an idle smoke, and their eyes more attentive to the show, it kindled inwardly, and ran round like a train, consuming within less than an hour the whole house to its very grounds.

This was the fatal period of that virtuous fabric, wherein yet nothing did perish but wood and straw, and a few forsaken cloaks; only one man had his breeches set on fire, that would perhaps have broiled him, if he had not by the benefit of a provident wit put it out with bottle ale."[68]

With his interests, there is every reason to believe that Wotton must have frequented the Globe theater when Shakespeare was presenting his plays there. *Othello* was written in 1604, the year that Wotton first went to Venice as the British ambassador. Then there is Wotton's meeting in the spring of 1638—the year before he died—with the then very young John Milton. Wotton had already read *Comus* but knew nothing about the author when Milton came to visit him at Eton bearing

---

[67] LPS, Vol. I, p. 314.
[68] LPS, Vol. II, pp. 32–33.

another copy. Milton was about to leave for Italy, and Wotton sent him a letter of advice that contains the famous sentences referring to the poetry, "Wherein I should much commend the tragical part if the lyrical did not ravish me with a certain Dorique [pastoral] delicacy in your songs and odes; whereunto I must plainly confess to have seen nothing parallel in our language, *ipsa mollities* [of comparable tenderness.] But I must not omit to tell you that I now only owe you thanks for intimating unto me (how modestly soever) the true artificer. " This appears to have been the first evaluation of Milton's poetry by anyone of distinction, and he never forgot it.

Other readers of these letters may find themselves drawn to Wotton's descriptions of people like Paolo Sarpi, often considered the greatest Venetian intellect of his age, or of the antique doges who ran the city's labyrinthine politics or of the spies who were employed by the ambassador to observe them. Because of my background I was especially interested in his references to contemporary science and scientists. We have already met Francis Bacon and Johannes Kepler. But there was also Robert Boyle, whose *The Sceptical Chemist*, which he published in 1661, turned alchemy into chemistry. Boyle was the 14th child of Wotton's friend the Earl of Cork. He was a child prodigy, and in 1635, at the age of eight, he was sent to Eton and taken in hand by Wotton. It was arranged that the young nobleman should be tutored in both violin playing and singing. He was also cured of a stammer. Boyle later wrote about Wotton's role with gratitude and great respect. But for me the most interesting had to do with Galileo. In the fall of 1608, the aforementioned Sarpi, who carried out an extensive correspondence all over Europe, learned that a Dutch lens-grinder named Hans Lipperhey had applied for a patent on what was the first telescope. Galileo heard about this in a visit to Venice in July of 1609 and proceeded to build one for himself. In fact his first telescope, which he demonstrated to Sarpi, was nine-power, three times as powerful as the original. By the end of the year he had built a 30-power telescope and had managed an appointment to the university in Venice. As was usual with Galileo, he had an argument over the terms and eventually ended up in Florence. But as soon as he had made the first telescope he began to make discoveries with it that transformed the accepted contemporary view of the universe. Wotton was in Venice at the time and, as the following letter shows, was practically beside himself with excitement. It was written to the Earl of Salisbury, Wotton's conduit to the king, on March 13, 1610. After some old business, Wotton writes, "Now touching on the occurrents of the present, I send herewith unto his Majesty the strangest piece of news (as I may justly call it) that he hath ever yet received from any part of the world; which is the annexed book (come abroad this very day of the Mathematical Professor at Padua [This is a reference to Galileo's *Siderius Nuncius, The Starry Messenger*, that had just been published in Venice. Galileo was still in the process of arranging his appointment at Florence. There is no indication, as far as I know, that James ever read this book], who by the help of an optical instrument (which both enlargeth and approximateth the object) invented first in Flanders, and bettered by himself, hath discovered four new planets rolling around the sphere of Jupiter. [These were the moons of Jupiter. Wotton reflects here the then-commonly held view that all the planets were attached to spheres that moved. It was Kepler who first introduced the idea of orbits. One wonders if he and Wotton discussed any of this at their meeting a decade later], besides many other unknown fixed stars; likewise the true cause of the *Via Lactea*, so long searched. [Galileo's telescope was able to resolve the individual stars in the Milky Way, thus showing that it was made up of stars.]; and lastly, that the moon is not spherical, but endued with many prominences, and which is of all the strangest, illuminated with the solar light by reflection from the body of the earth, as he seemeth to say. [When the moon appears only as a crescent, one can observe on the rest of its surface light reflected from Earth.] So as upon the whole subject he hath first overthrown all former astronomy—for we must have a new sphere to save appearances. ["Saving appearances" was the grand theme of planetary astronomy from the time of the Greeks. The idea was that planets being heavenly bodies made out of the "fifth essence" must move uniformly in circles. The problem was that this is not how planets actually move. Indeed, at times, they reverse course and move backwards— "retrograde motion. " To reconcile this with the circular motion assumption—to "save appearances" —circles were added to circles, making the whole thing a maze of complexity. Kepler did away with all of this by showing that the motion of the planets was in elliptical orbits around the Sun. Interestingly, neither here nor anywhere in his correspondence does Wotton mention Copernicus and the Sun-centered cosmology. Again, one wonders if any of this came up in his meeting with Kepler.] —and next all astrology. For the virtue of these new planets must needs vary the judicial part [presumably an astrological reference], and why may there not yet be more? These things I have been bold thus to discourse unto your Lordship, whereof here all corners are full. And the author runneth a fortune to be either exceeding famous or exceeding ridiculous. By the next ship your Lordship shall receive from me one of the above-named instruments, as it is bettered by this man."[69] I do not know what happened to this telescope, nor do I find in Wotton's correspondence any reference to Galileo's troubles with the church. Wotton does tell us of his meeting with Cardinal Roberto Bellarmine, who had such a significant role in Gallileo's encounter with the Inquisition.

Wotton left his last post in Venice in the fall of 1623. His prospects in England were not good. He was in ill-health and in serious financial need. Ambassadors were always owed money after their terms of service, and they were rarely paid. King James was at the end of his reign. He would die in 1625, and Wotton had no claim on his successor. To repair his finances—Wotton was actually arrested for debt—he wrote a popular little book on architecture, *The Elements of Architecture*, and began a difficult but ultimately successful campaign to become to Provost of Eton. He started at Eton in 1624 and remained in that post for the next 15 years. Toward the end he was often ill and, feeling that he did not have much longer to live, he prepared his will. The will is an interesting document since it is a kind of catalog of the paintings Wotton had accumulated, which he gave away. He had a very fine eye for art and sometimes bought paintings for other people as well as for

---

[69]LPS, Vol. I, p. 486.

himself. Wotton's paintings now hang in various museums. There are some paintings of him. My favorite is the one painted in 1620, which is now in the Bodleian Gallery in Oxford. It catches one by surprise. He does not look like a courtier but rather like an amused and hardy country squire—a Kentish squire—which is how he viewed himself. He looks like someone we all would have enjoyed knowing. In his letters he now and again quotes the toast *Lunga vita et bel morire*—to a long life and a dignified death. Wotton had both.

Part III
Linguists

Chapter 9
The Spencers of Althorp and Sir William Jones: A Love Story

One of our club at the Turk's head [Edward Gibbon, a fellow member of Samuel Johnson's literary club] has just published an History of the Decline of the Roman Empire, which I fancy you will read with pleasure: it is written in an elegant and easy style; but with very little verve or vigour.

—William Jones to the Viscount Althorp, later the 2nd Earl Spencer,
February 23, 1776

When I was at sea last August, on my voyage to this country, which I had long and ardently desired to visit, I found one evening, on inspecting the observations of the day, that India lay before us, and Persia on our left, whilst a breeze from Arabia blew nearly from our stern. A situation so pleasing in itself, and to me so new, could not fail to awaken a train of reflections in a mind, which had early been accustomed to contemplate with delight the eventful histories and agreeable fictions of this eastern world. It gave me inexpressible pleasure to find myself in the midst of so noble an amphitheatre, almost encircled by the vast regions of Asia, which has ever been esteemed the nurse of sciences, the inventress of delightful and useful arts, the scene of generous actions, fertile in the productions of human genius, abounding in natural wonders, and infinitely diversified in the forms of religion and government, in the laws, customs, and languages,as well as in the features and complexions of men.

—William Jones, 1784

In fact, I shall leave a country [India] where we have no royal court, no house of lords, no clergy with wealth or power, no taxes, no fear of robbers or fire, no snow and hard frosts followed by comfortless thaws, and no ice except what is made by art to supply our desserts; add to this, that I have twice as much money as I want, and am conscious of doing very great and extensive good to many millions of native Indians, who look up to me, not as their judge only, but as their legislator. Nevertheless, a man who has nearly closed the 47th year of his

J. Bernstein, *Physicists on Wall Street and Other Essays on Science and Society,*
© Springer Science+Business Media, LLC 2008

*age, and who sees younger men dying around him continually,
has a right to think of retirement in this life, and ought to think
chiefly of preparing himself for another; and my desire to pass
those years which it may please God to allot me in this distracted world, near those whom I love, will induce me to revisit
Europe, as soon as I can live in it, not as a Lucullus but as a
Cincinnatus, with perfect ease to myself and perfect independence on others, except perhaps on my ploughman and
gardener.*

—William Jones to the 2nd Earl Spencer, November 1793

In 1765, when he was 19 and a student at Oxford, William Jones was offered the position of tutor to the then seven-year-old Viscount Althorp, George John Spencer. There must have been an understanding that he would also instruct Spencer's sister, Georgiana, later the Duchess of Devonshire, who was a year older and to whom Jones taught writing. At the time of this offer Jones had never met the Spencers, nor had he met the man who had recommended him, Jonathan Shipley, the dean of Winchester and later the bishop of St. Asaph, who eventually became Jones's father-in-law. It appears that the dean had heard of Jones because of the brilliance he had displayed at Harrow, his public school. Among other things, Jones seems to have had a photographic memory. While he was at Harrow, the boys decided that they would like to act out *The Tempest* and, as there was no copy of the play availa- ble, Jones obligingly wrote one out from memory. He also mastered Greek, Latin, French, and Italian at Harrow. Late in his life Jones made a list of the languages he had learned. It reads:

Eight languages studied critically: English, Latin, French, Italian, Greek, Arabic, Persian, Sanscrit [sic]
Eight studied less perfectly, but all intelligible with a dictionary: Spanish, Portuguese, German, Runick [This may be a reference to Old Norse, one of the Runick languages], Hebrew, Bengali, Hindi, Turkish
Twelve studied least perfectly, but all attainable: Tibetian [sic], Pali [an early Indic dialect in which Buddhist canon is preserved], Phalavi [usually called Pehlavi or Middle Persian], Deri [a Persian dialect spoken in Afghanistan], Russian, Syriac [Aramaic], Ethiopic, Coptic, Welsh, Swedish, Dutch, Chinese

Being a tutor to a seven-year-old noble—a position slightly higher in the hierar- chy of a family like the Spencers than a fencing master—was certainly not Jones's first choice. But he needed the money. His father, also William, had died when Jones was only three. The senior Jones, the son of a small Welsh farmer, had attracted the attention of a local landowner, who got him a job in a London counting house. Jones never had a university education, but he taught himself enough math- ematics so that, after traveling to the West Indies, he got the curious job of teaching the subject on a British man-of-war. This presumably had to do with the principles of navigation, which became the subject of his first book, published in 1702, entitled *A New Compendium of the Whole Art of Navigation*. By this time he had returned to Britain, where he set himself up as a mathematics tutor in London. His clientele included members of several noble families, including Thomas Parker, Earl of Macclesfield, and his son George, who eventually became president of the Royal Society. Jones went to live at Shirburn Castle in Oxfordshire. Here he met Maria Nix, who was the daughter of a cabinet maker. They married and had three children, a son who died, a daughter Mary, and, finally Jones.

As a mathematician, the senior Jones would hardly be of any interest, but nonetheless he is well-known to historians of mathematics. In 1706 he published a book entitled *Synopsis palmariorum matheseos*, which was an elementary but contemporary mathematics textbook. He introduced the notation 'π' for the ratio of the circumference to the diameter of a circle, which is still with us to this day. But he also explained the rudiments of calculus, and whatever else mathematicians were then making use of. This book caught the attention of Isaac Newton, and Jones became a willing participant in the fight that was taking place between Newton and the German polymath Gottfried Wilhelm Leibniz, over which one of them had invented calculus and whether one of them had stolen the idea from the other. In 1708, Jones acquired a set of papers from one John Collins, which included an unpublished manuscript of Newton's dating from 1669; these papers showed that Newton had invented calculus independently. And considering his almost superhuman mathematical powers, this shouldn't have left any doubts. Nonetheless, Newton, who was relentless when challenged like this, had a Royal Society committee set up to officially rule on the matter. Jones was a member. It is not surprising that the committee ruled in favor of Newton. Leibniz died in relative obscurity in 1716, with Newton still pursuing him beyond the grave. Jones became a friend of Newton's and a fellow of the Royal Society and eventually its vice president. When he died in 1749, he left a massive assortment of manuscripts, some of which were his own and some Newton's, that scholars are still disentangling. He had intended to publish them. Later on, his son learned enough mathematics to begin the job of completing his father's work, something that he never got very far with.

There seems to have been enough of an estate left by the senior Jones so that the family, while by no means wealthy, was able to send Jones to Harrow and then Oxford. Jones's mother was a remarkably forceful woman who was determined that her son have the best possible education. Indeed, in 1765, she moved to Oxford with Jones's sister so as to be near him, at least when he was resident in college. She might have been more protective than the average mother because of Jones's childhood history. When he was four, in attempting to scrape some soot from a chimney, he fell into the fire and was badly burned. A little later some servants trying to dress him inadvertently stuck a hook-fastener in his eye. He had trouble with his vision for the rest of his life. At age nine he broke his thigh in an accident that kept him bedridden for a year. This disability made athletics difficult for him at Harrow, which is one reason why he focused so intensely on academics.

Even though he was still at Oxford, after he became George John's tutor he often stayed with the Spencer family. There was some ambiguity as to where the Spencer residence was. In addition to the family country seat, Althorp Park, about 100 miles north of London in Northhamptonshire, the Spencers, one of the wealthiest families in Britain, had four other homes. These included a villa in Wimbledon, a hunting lodge in Pytchley, and a magnificent newly constructed mansion in London in St. James. It is clear from his correspondence that Jones was often alone with George John and Georgiana while their parents were in one of the other houses. Indeed, from the beginning, Jones had a special relationship with the children. In George John's case it went beyond being simply being his tutor. He became his friend, possibly his best friend. He was also close to Georgiana, but as she grew into adulthood her life became so chaotic that Jones stayed out of it. Indeed, one wonders how much he knew about it. There is certainly no hint in any of the letters. It was quite different with George John. Even when he was in India, and it could take a year for an exchange of letters, Jones's letters—making due allowance for their British reserve—are the sort of letters one could only write to someone whom one both loved and had the deepest confidence in. The tone is set from the very first letter, which was written in November of 1768. Keep in mind that the letter is being written to a boy of ten. Note the salutation. Jones never called either of these children by their first names. It is written from Oxford. It reads:

My dear Lord Althorp

I took my degree yesterday morning, which was sooner than I expected, but as I have ordered my post-chaise for Sunday, and have not taken leave of all my friends yet, I fancy I shall not have the pleasure of seeing you till then. When we meet I will give you an account of my taking the degree; which was the most ridiculous ceremony imaginable. But at first it was rather serious, for I was extremely distressed to find nine Batchelors of Arts to *Scio* for me. Methinks I see you puzzling about the meaning of the word *Scio*. The case is this. It was absolutely necessary for me (as it is for every one that takes a degree) to get nine batchelors to recommend me to the convocation, in these words Scio Jones esse habilem et idoneum ad hunc gradum, I *know* that Jones is fit for this degree. But as Custom had brought it about, they leave out the latter part of the sentence and only say *Scio Jones*, whence they have formed the verb to *Scio* for such a person. I was a long time before I could get, any nine Scio's but at last they came and everything went on well. It is with infinite pleasure that I hear, I shall not be obliged to reside often at Oxford for my Master's degree; so that I shall have your company for many years to come without interruption, and when I come to town we will apply ourselves diligently to parsing Greek and Latin to verse, and to Latin composition, and then, away we go to the land of play, I mean school. I have sent you a Greek tree which you may copy out if you will, it contains all the first persons of all the tenses in Greek and shows you at one view how they are formed. I passed a very merry Evening last night at my mothers with five or six young men of my acquaintance; we played at dumb crambo, at Bouts Rimès, and all our old games; the bout rimè that raised the greatest laugh was my first; the words that were given me were winning, groaning, growling, sinning, moaning, howling, hunting—which I filled up thus in ridicule of the modern songs and pieces of poetry for musick:

Symphony) Ye lads that love winning,
Ye saints that love groaning,
Ye scolds that love growling,
Ye sots, that love grunting,
Ye monks, that love sinning,
Ye maids, that love moaning,
Ye bucks, that love howling,
Ye squires, that love hunting
Air.) Sing, winning and groaning,
Sing growling and grunting,
Sing sinning and moaning
Sing howling and hunting-
With hunting, and howling, & moaning, & sinning,
With grunting & growling, & groaning & winning

But I think I have made my letter long enough, or perhaps too long for what it contains. Therefore farewell, and believe me to be

Your ever affectionate friend
W. Jones

Some of Jones's letters, especially to Continental scholars, were written in Latin and often contained portions in four or five different languages. His later letters to George John often contained two or three.

In addition to languages and letters, Jones taught the children to play chess, something he had learned at a young age and played all his life. When he was a student at Harrow he wrote a poem, *Caissa, or The Game of Chess*, which one still finds quoted in anthologies of chess literature. The pieces and moves are reduced to human form and the battle is real. Of the king he writes:

High in the midst the reverend kings appear,
And o'er the rest their pearly scepters rear:
One solemn step, majestically slow,
They gravely move, and shun the dangerous foe;
If e'er they call, the watchful subjects spring.
And die with rapture if they save their king:
On him the glory of the day depends,
He once imprison'd, all the conflict ends.

When he got to India one of Jones's many research projects was to investigate the origins of chess.

From the beginning, Jones did not think of himself, when it came to the Spencer children, as a mere schoolmaster. It was, certainly in the case of George John, more as an Aristotle shaping a future Alexander. In this, Jones was not entirely misguided, since George John grew up to be one of the most distinguished men in Britain. After Harrow, George John went to Cambridge and then, in 1780, entered Parliament. In 1783, after his father's death, he became the second Earl Spencer, which qualified him for the House of Lords. He then occupied several important governmental posts, which culminated in his being appointed first lord of the admiralty in 1794, an office which he held for six years—six of the greatest years in the history of the British Navy. One of the prerequisites of that job was naming the first wooden gunboat in the fleet, which was launched in 1797. He named it "Blazer," after one of the dogs in his foxhound pack. He became home secretary from 1806 to 1807, after which he retired to private life. One of the things he devoted himself to was amassing what was then the finest private library in Europe. In the meantime, the rest of his family was disintegrating. In 1774, Georgiana married William, fifth Duke of Devonshire, a man of enormous wealth who seems to have had less interest in his extraordinarily beautiful and talented wife than he did in his foxhounds. The result was predictable; a litany of extramarital affairs and illegitimate children. Added to this was Georgiana's addiction to gambling, which was shared by both her mother and father and her sister Henrietta, who was four years younger and had an equally disastrous marriage. Georgiana was never out of debt. Some of her creditors were paid off by her husband and some by George John, but many never got paid. Her mother and father also lost huge sums of money gambling, and after the first earl, George John's father, died in 1783, it was left up to his son to pay the debts on the estate and to provide for his mother. His own marriage to Lavinia Bingham, the eldest daughter of the first earl of Lucan in 1781, appears to have been a stable one. Jones attended the wedding and produced an ode *The Muse Recalled*, a few lines of which give the general flavor:

> While thou, by list'ning crowds approv'd,
> Lov'd by the Muse and by the poet lov'd
> ALTHORP shouldst emulate the fame
> Of Roman patriots and th' Athenian name....

It seems reasonable to credit Jones with playing a crucial role in the formation of George John's character, which was so different from both his parents and siblings. But their relationship almost came to an abrupt end. In 1770, Jones was prepared to sever his connections with the Spencers entirely. There were several reasons. In the first place, Jones had decided to leave Oxford to study law, although he had been awarded a fellowship at Oxford and had already begun to acquire a reputation as an Orientalist. In fact, four years earlier, through his connections with the Spencers, he had met the duke of Grafton, then the head of the treasury, who had offered him the position of Interpreter for Eastern languages, which he turned down. But two years later an undersecretary to the duke made him a second proposition. A Persian manuscript—a history of Nadir Shah—had been brought to England in 1768 by the king of Denmark, Christian VII. He wanted it translated into French. Jones was at first not very enthusiastic but eventually took on the job. *L'Histoire de Nader Shah*, which also included some translations of Persian verse, was published at Jones's expense in 1770, with several copies going to the king. With his small fellowship he certainly could have eked out a bachelor's existence at Oxford. In 1766, he met Anna Maria Shipley, the eldest daughter of the aforementioned Jonathan Shipley. She was a frequent visitor to the Spencers. In fact, she was related to Lady Spencer. Jones might well have proposed to her, but he did not have the money to support a wife and would not take any money from her parents. The only choice was for him to find a more lucrative profession and wait, which they did—for 16 years.

George John matriculated at Harrow in June of 1769. Jones accompanied him, again as a tutor. They shared lodgings. But after the summer George John developed shortness of breath, and his mother decided to delay his return to school for a year. This left Jones free to continue his work on his translation. The Spencers assumed that he would stay on as a part-time private tutor to their son, but Jones had other ideas. On October 10, 1770, Jones wrote a letter to Lady Spencer of such temerity that it could well have brought his connection to the Spencer family to an end. Here is what he wrote in part. It is in response to the suggestion that George John should split his education between private tutoring and public school and so divide his "hours." "I do not want his hours only," Jones writes, "but his months and years. I wish to be with him at meals, at our amusements, in our hours of rest as well as of study that I may give his thoughts a right turn upon every subject, and strengthen his mind while I seem to relax it. I, therefore, beg leave to propose that it be maturely considered whether Lord Althorp's health will permit him without any risk to pass the three or four best years at Harrow without interruption; if not, that a private house be formed upon that footing, where I may pursue my plan without being turned from it by the avocations of a family. I hope he may come to the University when he is sixteen, and till that age not lose a single hour." Needless to say, both Lord and Lady Spencer responded to this suggestion with a resounding no. By December, George John had returned to Harrow, and Jones's role as his private tutor was over, but not their relationship which, if anything, deepened.

During George John's years at Harrow and Cambridge, Jones's role developed into being almost that of a surrogate older brother. His letters to the Spencers reflect that. To Lady Spencer he writes in February of 1773, "There are *three* authors, with whom I hope, one day or other, to bring him acquainted, Euclid, Locke and Blackstone: the first will open up to him the principles of all natural knowledge, the second will show him the nature and extent of human reason, and the third will offer to him a specimen of that reason reduced to practice in the Admirable Laws of our country; and here I cannot help observing, that, with how much pain and reluctance soever I separated myself from him two years ago, yet, without that separation, I should never have been able to make my studies so useful to him, as they may now perhaps be; since I should never have applied myself to the sciences, an especially to that profession [the law], in which I am now ambitious of excelling." In December of 1774, he writes to George John, "What company do you have at Althorp? What Musick? What new dances? What adventures have you met with? Are you able to play in concert with your violoncello? That is another accomplishment, a little of which, like salt, gives relish to life, but too much spoils the whole dish.—I hope Lord Spencer has got rid of his gout, and that he will have a fine season for hunting. I suppose the duke and duchess of Devonshire [Georgiana and her husband] are with you. As you were so good a lord chamberlain when I had the pleasure of seeing Lady Spencer, I shall desire you to present me to the duchess, if we meet in town next spring. I fear I ought to have made my bow before, but I did not know the étiquette, and a philosopher among courtiers, who flock together on such occasions, is like a lark or a nightingale in a *ménagerie* of peacocks or Indic pheasants." In reading the last two sentences, keep in mind that this was the same duchess, Georgiana, that Jones had taught to write! It is characteristic of Jones, at least until he got to India, that he always felt himself to be a "philosopher among courtiers," someone who was looking on from the outside.

In the years immediately after he left the employ of the Spencers, Jones had a dual career. On the one hand he was studying law at the Middle Temple in London. He began practicing law in 1774. On the other he was still continuing his work in Oriental languages and literature. Indeed, in 1771, he published his very influential *A Grammar of the Persian Language*, which included some of the first translations into English of important Persian poetry. This was an era of sobriquets, and Jones became known thereafter as "Persian Jones" or "Oriental Jones." Samuel Johnson was, needless to say, known as "Dictionary Johnson." He was the guiding spirit of the Literary Club, which met once a week at the Turk's Head to dine and talk until early morning. The members included men such as Joshua Reynolds, Edmund Burke, Oliver Goldsmith, and Edward Gibbon. Jones had known Reynolds earlier. Reynolds had done a portrait of him. This portrait, which was painted when Jones was in his twenties, shows a full-faced amiable looking young man. When one compares this to the portrait made in India by the artist A. W. Devis when Jones was 47, after the rigors of Indian life and illness had taken their toll, it is difficult to believe that they are the same man.

Jones began being invited as a guest at the Literary Club before he became a member in 1773, a great honor for such a young man. Later on he was able to persuade George John Spencer to become a candidate for membership and to see him successfully elected. In a letter written to George John on November 30, 1778, he describes the members. Here are a few entries:

Burke, the pleasantest companion in the world...*Fox* [Charles James], of great talents both natural and acquired; *Gibbon*, an elegant writer, not without wit in conversation...*Garrick* [David of the theater], whom all Europe knows, *Sheridan* [Richard Brinsley, the playwright], a sprightly young fellow with a fine comick genius; very little older than yourself; ... *Johnson*, the best scholar of his age; ...*Smith* [Adam], author of a great work on the wealth of nations; *Reynolds*, a great artist and fine writer on his art; *Boswell*, of Corsica, a good natured odd fellow: ...R. *Chambers*, an India judge; and W. *Jones*, likely to be his colleague.

The last entry shows that Jones has already set his sights on India.

Jones's law practice was divided. On the one hand he had clients in London, where he now lived when he was not at Oxford, and on the other, he made the circuit. This consisted of going on horseback or carriage to various towns in Britain and Wales. He was also doing the odd legal work for Lady Spencer. In January of 1776, George John enrolled at Trinity College in Cambridge, and, on the occasion, Jones wrote an affectionate letter of advice which begins.

"My dearest lord"

Such is my heart, that it has very often felt the highest gratification's and pleasures; but a pleasure greater than that which your last letter gave me, my mind never yet received. My nature is free and open, warm in friendship, anxious (perhaps to an extreme) for the welfare of my friends, consequently liable to be depressed by the disappointment of my wishes for their glory and happiness, nor less apt to be transported with joy at the completion of my hopes. This joy I now feel to a great degree; and, as you attracted my regard and affection at the age of seven years (when I was twelve years older) which affection has been continually increasing for ten years more, I concentrated all my hopes of perfect satisfaction in this life in the prospect of your becoming a virtuous and able man, a lover of your country and a benefactor of mankind. The morning of your life, spent at a publick school, has answered my warmest expectations; and your letter, which I have just received, has fixed the crown of pleasure and comfort on my head; nor shall I envy all the crowned heads on earth, if you advance in the same manner towards the noon of your mortal state. The University is the next part of your career, and the sentiments which you elegantly express about the true ends of Learning (in which you know how perfectly we agree) give me full assurance that you will finish it with equal honour, I approve, my dear lord, I approve with my whole heart, the opinions contained in your letter. Persist in the study of our history according to your method, continue your taste for musick, regulated by your good sense, and enjoy at proper seasons the pleasure of manly exercise and social recreations; but fix your mind intent, as an archer fixes his eye on the mark, on the grand object of life, the benefit of our country and of all human species. Further advice on this head will now, I find, be superfluous; I have perfect confidence in you: steer right onward, and your ship will reach the destined haven without danger. Excuse me, however, if I say that I cannot help being a little anxious about your Academical course at Cambridge: yet why should I be anxious? You will find the *young* men, as they are at all Universities, too much addicted to the pursuit of pleasures: but *frivolous* pleasures you will despise; and those which are not wholly innocent, you will detest. The *old* men you will find, as they are in all places, too attentive to their *interests*, and consequently servile to men of rank in hopes of promotion by their help; but I have such reliance on your good understanding, that I am sure you will be in no danger from abject flattery, which you cannot but contemn. I exhort you however to beware of flattery under the mask of friendship, for it is very insinuating, and few minds are strong enough to resist it. Yours will, I know, resist it; and with a little caution of this kind, I promise you great satisfaction comfort and improvement in your situation at college."

Jones's legal practice was never as lucrative as he had hoped for. Perhaps he was distracted by his Oriental studies, which continued and took a good deal of his time. Perhaps he was too blunt. Perhaps his politics were too radical. He shared with the Spencers their Whig traditions, but he was further out on the spectrum than they were. Jones believed very early that the British should give the American colonies their freedom, something that it took a few years before the mainstream Whigs like Burke came to accept. (Dr. Johnson was a dyed in the wool Tory but, at the Literary Club, this was no impediment to membership. Jones and Johnson did become somewhat estranged over Johnson's underestimation of Milton, who was one of Jones's poetic gods.) Jones became a good friend of Benjamin Franklin, and if he hadn't gone to India he might well have settled in America. He had begun to make plans for such a migration. In 1780, Jones made a run for Parliament as a representative from Oxford. Candidates were not allowed to solicit votes directly on their own behalf but could enlist others to do so. Georgiana, who was deep into Whig politics—she had a very close friendship with Fox—tried to help, but to no avail. Jones withdrew from the race when it became clear that he had no chance to win. In the meantime he had set his sights on a judgeship in India.

To understand Jones's goal we must briefly describe how these judgeships came to be created. By the early 1770s, the British East India Company, which after the Bank of England was the most important financial institution in Britain, had gotten itself into serious financial difficulties. Indeed, it was on its way to going bankrupt. Part of this had to do with corruption on the part of its employees in India, who siphoned off huge sums of money. Part had to do with the fact that the Company had to pay large mercenary armies to maintain its slender foothold in India, and part had to do with the vagaries of an agrarian economy. A failed monsoon could produce a general famine. The government had been anxious to exert some measure of control over the Company, and these financial exigencies gave it the leverage it needed. As a condition of a loan that would allow the Company to manage its debts, the Tory government of Frederick Lord North managed to pass two parliamentary acts. One was the so-called Regulating Act of 1773, which created the position of governor-general, along with a four-man council, which would rule the three British colonies; Bombay, Madras, and Bengal. The first governor-general was Warren Hastings, something that would turn out to be of great importance to Jones. Parallel to the Regulating Act was the Judicature Act, which created a Supreme Court located in Calcutta. This court was the final arbiter of all legal matters for the natives of Calcutta and for British subjects and company employees elsewhere. The court would consist of a chief justice and three so-called puisne judges, who were to be paid 6,000 pounds a year by the company—a very high salary.

The first chief justice was Sir Elijah Impey, who had been a schoolmate of Hastings at Westminster. It was Jones's hope to replace one of the puisne judges if one of them should retire or die. As far as he was concerned, this was his ticket to the future. He would finally have enough money so that he could marry and, in a few years, he thought, save enough from his salary so that he could retire to England, perhaps even buy a seat in Parliament, and live the rest of his life in comfort. Indeed, he reasoned that he could save some 4,000 pounds a year and that he would need about 30,000 to retire, so that with the interest compounding it would not take too many years to reach his goal. And besides, for an Orientalist, India was the Promised Land. The problem was getting appointed.

In 1777, one of the original appointees, Stephen Caesar Lemaistre, died and it looked momentarily as if Jones might actually have a chance of achieving his ambition at a fairly early date. But there were obstacles. In the first place, no barrister could be appointed a judge unless he had been a barrister for at least five years. As Jones had begun his career in 1774, there were a few years to wait. But, more fundamentally, the Indian judgeship was a political appointment, which intertwined the relationship of the government with the company and that of any candidate with the government. At this time it was not entirely clear that the Bengal Supreme Court would endure as an institution. Moreover, Jones's radical views about such things as the independence of the American colonies had gotten him into a certain amount of trouble with the establishment. In particular, Edward Thurlow, who was the Lord Chancellor during this period, the highest legal authority in the country, was a Tory who supported the war. Whatever Jones's other qualifications were, this was enough to block his appointment. In March of 1780, Jones wrote a letter to George John which summarizes his understanding of the situation. He writes, "The final disagreement between the minister [Lord North] and the India-company will keep me longer in suspense about the judgeship: I should like the appointment: it would coincide with all the variety of my studies and inclinations: my system of temperance would preserve my health in any climate under heaven: I should be certain of returning (if I did return) with a very handsome fortune, and might be in parliament before the age of forty, with a perfect knowledge of India, which is now become an object of infinite importance to every British statesman. After all I cannot relish this state of suspense: the general election approaches, a time when every lawyer wishes to step forward on the scene of action: I can ill brook the necessity of remaining so long at my age without taking a part in the affairs of my country.... On the whole, if nothing be determined, as to my promotion before the end of this session I shall spend a few weeks in deliberating, and consulting my best friends, whether it will be wiser for me to renounce all idea of the judgeship, and to enter boldly on my political career at a time, when malice itself cannot ascribe my words or actions to any motive of resentment. In fact I neither am, nor can justly be, angry with Lord North. Allowance must be made for the singularity of his disposition and manners, which is inconvenient to his best friends. If I were to ride an elephant, could I with justice be angry because I could not travel so fast as on horseback? No: it is the nature of the animal—Besides, I clearly perceive the minister's delicate situation in respect for the India-company. He could not wisely have supplied the vacancy, whilst an abolition of the court was in agitation. —I therefore have no reason to complain; but as I never act by halves and the part which I shall take in my country both in speaking and writing will be very determined, I wish to prepare my way as to put it out of any man's power to say justly that I act from disappointment. You, my friend, who have so long known me will bear witness that my sentiments have been uniform; and next to the love of liberty, I desire to be thought uniform in nothing so much as in friendship for you. Farewell, again and again."

North was notorious for dealing with difficult situations by delaying decisions on them. One of the decisions he was delaying was leaving the government because of his growing concern over the war. On November 25, 1781, General Charles Cornwallis surrendered to Washington at Yorktown, and for all practical purposes the war was over. By the spring of 1782, North had resigned from the government, which fell. By the summer a new, and as it turned out, short-lived government was put in place with the Earl of Shelburne as its prime minister. Sensing a window of opportunity, Jones addressed his concerns about the judgeship in Bengal directly to Shelburne. This time it worked, and in February of 1783, as one of his last acts as prime minister, Shelburne recommended to the king that Jones be appointed. The king wrote a strong letter to Thurlow on the first of March proposing Jones's appointment as a favor to Shelburne, and on the fourth of March the appointment was publicly announced. Jones was granted a knighthood, and on April 8 he and Anna Maria Shipley were married. A few days later the newlyweds sailed for India on the frigate *Crocodile* and a new chapter in Jones's life had opened.

Jones's letters from India are among his finest. The ones to George John began on the voyage out. On the April 22 he writes, "The Crocodile is almost new, and, though small, an excellent sea-boat; the captain intelligent and experienced, eager to oblige, desirous and capable of entertaining; the officers men of agreeable manners and good sense. My daily studies are now, what they will be for six years to come, Persian and Law, and whatever relates to India; my recreation, chess; my exercise walking on deck an hour before dinner; but my great delight is the sweet society and conversation of Anna Maria, whose health and spirits are really wonderful in a situation so new to her and by no means pleasing in itself. The motion of the ship obliges me to lay down my pen. Farewell my lord; I will write again if I am able, from Madeira." The voyage lasted until September 25, when the ship landed in Calcutta, an average time for a passage to India.

Jones's arrival had been highly anticipated, above all by Hastings, whose knowledge of India was unsurpassed. Hastings was not a professional linguist but he had learned Persian, Urdu, Hindustani, and Bengali. He needed these in his work. He had never learned Sanskrit but had sponsored a young employee of the company, Charles Wilkins, in his attempt to master the language with the aid of the native pandits. Eventually Wilkins translated the *Bhagavad-Gita* into English for which Hastings supplied the preface. He had also encouraged another of his charges, Nathaniel Halhed, to translate a compendium of Hindu law—the *Code of Gentoo Law*—and to produce a Bengali grammar that used Bengali-type characters Wilkins had created. Hastings and Jones became very close friends, although their overlap in India lasted only 16 months. After Edmund Burke led the impeachment proceedings against Hastings, which began in 1786, Jones had by then ended his friendship with Burke.

Jones found his life in India very demanding. A few years later, when it had developed into a routine, he described it in a letter to George John, who by this time had inherited his father's title and had become the second earl of Spencer. Jones writes, "Would you like to know how I pass the day? I will give you a sketch of my employments during the term and sittings, and during the vacation. In term time I rise an hour before the sun, and walk from my garden to the fort, about three miles: thence I go in a palanquin [a covered platform usually carried by four men] to the Court-house, where cold bathing, dressing and breakfast take up to an hour; so that by seven I am ready for my Pandit, with whom I read Sanscrit [sic]; at eight come a Persian and Arab alternately with whom I read till nine except on Saturday, when I give instructions to my Mogul secretary on my correspondence with the Musilman scholars. At nine come the attornies with affidavits: I am then robed and ready for court, where I sit on the bench, one day with another, five hours. At three I dress and dine; and till near sunset, am at the service of my friends, who chuse to dine with me. When the sun is sunk in the Ganges, we drive to the Gardens either in our post-chaise, or Anna's phaeton drawn by a pair of beautiful Nepal horses. After tea time we read; and never sit up, if we can avoid it, after ten. But, for four months in the year I must sit three evenings a week as a justice of the peace. In the vacation, when we are at our villa in summer, and at our cottage in autumn, I rise when I happen to wake, and after bathing and coffee, read *Sanscrit*, till about eleven, when I read English or Italian to A.M. [Anna Maria] for an hour, (and on Sunday's, books of Theology) after which I finish the rest of my day's task till dinner- time. At sunset I walk two or three miles, while she is carried by four men in a chair, called a *Ta⁻nja⁻n* pronounced *Tonjong*, and used in the Eastern Peninsula, from which she brought it. On our return, we pass our evenings as we generally do in term-time. Your will judge therefore, whether I wish to change this calm course of life for the house of commons, from which I should return three or four nights in seven with *despair* & the head-ach. Farewell!"

The amount of work Jones managed to produce during this period almost defies comprehension. In 1784, not long after he arrived in India, Jones created the Asiatic Society. He modeled it after the Royal Society in Britain of which he had been a member since 1772. It was to be a place where the accumulated knowledge about Asia could be discussed and recorded in the society's journal. The presidency was offered to Hastings, who declined, so that it devolved to Jones, who did all the editorial work and much of the writing for the journal. He suggested the possibility that learned Indians might be invited to join at some time. He described the society to George John in a letter he wrote in August of 1787. "Do not," he wrote, "raise too high your expectations of entertainment or instruction from the Transactions of our Society, which they print so slowly (the Government constantly using the Press for Orders, Regulations &c.) that, I left only eight sheets printed, though we have mate-rial for two volumes in Quarto. It is not here, as in Europe, where many are scholars and Philosophers professedly, without any other pursuit: here every member of our Society is a man of business, occupied in his respective line of revenue, commerce, law, medicine, military affairs, and so forth: his leisure must be allotted, in great part, to the care of his health, even if pleasure engage no share of it. What part of it remains, then for literature? Instead, therefore, of being suprized, that we have done so little, the world if they are candid, will wonder, that we have done so much. *Sanscrit* literature is, indeed, a new world: the language (which I begin to speak with ease), is the *Latin* of *India*, and a sister of Latin & Greek. In Sanscrit are written half a million of Stanzas on sacred history & literature, Epick and Lyrick poems innu-merable, and (what is wonderful) Tragedies & Comedies not to be counted, above 2000 years old, besides works on Law (my great object), on Medicine, on Theology, on Arithmetick, on Ethicks, and so on to infinity. Farewell."

Jones's fairly casual remark in this letter about the family resemblance between Sanskrit and European languages like Latin and Greek is a precursor to one of his most important observations, in which he stated in a paragraph in the third of the eleven annual reports he wrote for the Asiatic Society. He asserted that there was a group of languages which included Sanskrit and Greek—which are now known as Indo-European—which had evolved from a vanished and now unspoken proto-language. This language is now known to linguists as Proto-Indo-European—PIE. It appears to have arisen in Central Asia thousands of years before the Christian era and had no written counterpart. In 1786 Jones delivered an address. I will quote from it in the next essay, which anticipated the modern study of historical linguistics.

By this time, George John had become Jones's confidant in every way. He was handling Jones's financial affairs in Britain, helping him save for his retirement. He was one of the few people who knew of Jones's intention to retire. Jones had not made it widely known because he was concerned that if he did, so many people would want the job that he might be forced to retire before he was ready. In his letters he does not discuss at length his own fragile health, apart from his eyes, which always gave him trouble. His concern is for Anna Maria, who did not do well in the Indian climate and was constantly ill. Finally, in November of 1793, he managed to persuade her to return to England with the intention of following her as soon as he could finish the major work he had now started, the translation into English from Sanskrit of a digest of Hindu law. He never lived to see it completed. On the morning of the April 27, 1794, John Shore, who was then the Governor-General and who, as Lord Teignmouth, was to write the first biography of Jones, was summoned to Jones's house, where he was told that Jones was dying. By the time he got there, Jones was already dead. He was 47. The stated cause of his death was an inflammation of his liver. The real cause was India, which had supplied the oxygen in which his fragile candle had burnt too fast.

The pall that Jones's death cast on the community in Calcutta was enormous. It is well described in the memoirs of William Hickey, a cynical and worldly lawyer who practiced in Calcutta. There is nothing cynical in what Hickey wrote:

"The death of this enlightened and very learned man was properly felt to be a public calamity. The event was equally lamented by the natives as by Europeans, for all felt and acknowledged his extraordinary talents and his unblemished integrity as a Judge. Luckily his lady had left India (from the climate not agreeing with her) previous to the melancholy event. Had she been on the spot, such was the nature of her attachment to her husband, that in all probability the being present at his last moments would have proved fatal to herself. My absence from the Presidency prevented me from being at Sir William's funeral, which was attended by nearly the whole of the Settlement. He died after only a few hours illness, and was according to the custom of India…buried the same day."

Anna Maria did not learn of her husband's death until six months later. She shared her grief with the Spencers. Georgiana wrote a poem the last lines of which read:

Admir'd and valued in a distant land,
His gentle manners all affection won:
The prostrate Hindu own'd his fostering hand,
And Science mark'd him for her fav'rite son.
Regret and praise the general voice bestows,
And public sorrows with domestic blend;
But deeper yet must be the grief of those,
Who, while the sage they honor'd, lov'd the friend.

Chapter 10
All That Glitters

Visitors to the Mycenaean Room in the National Archaeological Museum in Athens are immediately struck by a tall glass case full of gold—Schliemann's gold. These are the wonderful gold artifacts that the German archaeologist Heinrich Schliemann found in the summer of 1876, in the "shaft graves," which are deep tubular burial sites in the palace in Mycenae on the Peloponnese in southern Greece. There are gold death masks, leaves, cups and statuettes, and jewelry. They are some 3,500 years old. They dazzle the eye. You can't miss them where they sit in the center of the room. However, few visitors bother to look into an unobtrusive glass case that stands off to the side, next to a wall. Those who do usually glance briefly at its contents and then move on. It is not hard to understand why. The case is filled with what, at first sight, appear to be oversized "skipping stones," the kind you look for at the edge of a pond to try to make hop over the surface of the water. These flat "skipping stones," which are clearly made of clay, are either rectangular or oval. They are at most about 6 inches in any dimension. Many are much smaller. All of them could be held in the palm of one hand. They are in various shades of brown, ranging from tawny to almost black. They look as if they had been in a fire.

If you look closer you will see what at first look like the tracks of agitated birds that might have landed on the clay while it was still wet. But then something else strikes you. The tablets have been carefully lined up horizontally. They look like pages from a notebook.

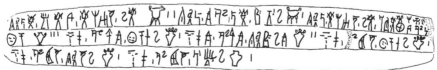

6 Pylos tablet Ta 641, showing tripod-cauldrons

Moreover, the scrawls, whatever they are, are carefully arranged along these lines. Furthermore, they often do not go all the way to the right-hand margin. Sometimes they stop in the middle and then continue on the next line, starting from the left. In publishing lingo, one would say that the left margin had been "justified."

J. Bernstein, *Physicists on Wall Street and Other Essays on Science and Society*,
© Springer Science+Business Media, LLC 2008

Whatever these symbols represent, they have clearly been set down in a left-right sense. But some symbols recur often. It eventually dawns on you that they must represent a number system. If one studies enough of them, it is not difficult to work out what it is. A short vertical line represents a unit, so that two such lines must stand for the number '2.' But after nine such lines what might have been a tenth vertical line is replaced by a single horizontal line, clearly standing for the number '10.' Nine of these are replaced by a circle standing for '100,' and nine circles are replaced by a circle with a dot in the middle, the '1,000,' and nine of these are replaced by a circle with a dot in the middle with rays coming out of its perimeter, the '10,000.' This is as high as the number system—which is evidently decimal—goes.

If you look at the tablets even more carefully, you will see a few ideographic signs, pictures that represent various objects. There is one that looks like a wheel with four spokes; another looks like a vessel with various handles and a tripod at its base. There are swords and arrows, symbols for men and women, and one that represents a chariot. A natural deduction is that these tablets are inventories of some sort.

This was the conclusion of Sir Arthur Evans, the self-taught archaeologist who first dug up tablets like these in 1900, though not the ones in this case, which came from the Greek mainland.[70]

Schliemann was also self-taught and, like Evans, wealthy. Schliemann's money came from various businesses he had created, while Evans money came from his family. The tablets Evans discovered were at Knossos, a palace complex near Herakleion on the island of Crete. Eventually, Evans purchased the site and reconstructed the palace. It was such a marvel of engineering and architecture that Evans was sure that any civilization with such gifts must also have a written language. Moreover, a few years earlier in Athens, he had come across small seal-stones with various kinds of hieroglyphics on them. He was told that they came from Crete. Therefore he was looking for some sort of writing when he began excavating. Indeed, he found more samples of the seal-stone hieroglyphics on Crete. The stones, which were known as "milk stones," were used as amulets by nursing mothers. But he also found tablets with two distinct types of linear writing. The writing that seemed more primitive he called Linear A, and the more sophisticated type, such as that found on the tablets in the Archeological Museum, he called Linear B. As he discovered more tablets—eventually in the thousands—he made all the deductions about Linear B that we have described, but that was almost as far as he could get. He assumed that Linear B was an evolved form of Linear A, but he could not decipher either one, and neither could anyone else for nearly a half a century.[71] Enter Michael Ventris.

---

[70] There is a fine new biography of Evans, *Minotaur*, by Joseph MacGillivray, Hill & Wang, New York, 2000. MacGillivray, an archaeologist living in Crete, notes that Evans was less than 5 feet tall. This is almost impossible to believe until one looks a photograph of him taken as an adult with a group of still active adult fellow boy scouts. He looks miniscule.

[71] The hieroglyphics have never been deciphered. They bear some resemblance to Egyptian hieroglyphics from which they may have been derived. The Egyptian hieroglyphics have been deciphered, but the underlying language is not that of Linear A, whose language remains unknown. For a good discussion of these matters see *The Story of Archaelogical Decipherment* by Maurice Pope, Scribners, New York, 1975.

Michael Ventris was born to a Cambridgeshire family on July 12, 1922.[72] His father was an Army officer in India and his mother half-Polish. They were well-to-do, sufficiently so, so that Ventris was sent to a private school at Gstaad in Switzerland for his elementary education, which was taught in French and German. He also mastered the local Swiss-German dialect. At the age of six he taught himself Polish. His ability to learn languages was remarkable. Some years later he learned Swedish on a visit of a few weeks.

Ventris returned to England from Switzerland and won a scholarship to the Stowe school where he studied Greek as well. He had now decided to become an architect, so he skipped college and enrolled directly in the Architectural Association School in London. Then the war came, and he enlisted in the R.A.F. Instead of becoming a pilot, he became a navigator in a bomber squadron. Navigation, he thought, was more stimulating intellectually than piloting. In 1948, he graduated from his architectural school with honors and began a career in architecture that involved among other things the design of schools for the Ministry of Education.

When Ventris was 14 he went to the Burlington House in London to see an exhibition marking the 50th anniversary of the British School of Archeology in Athens. It featured a talk by Evans in which the mystery of Linear B was discussed. Ventris was so intrigued that he decided then and there to try to decipher the language. When he was 18, concealing his age, Ventris managed to publish an article, "Introducing the Minoan Language," in the *American Journal of Archeology*. In it he addressed the fundamental question concerning Linear B, of what language was it the written counterpart. Evans had very definite ideas about this. He called the underlying language 'Minoan.' For him, Knossos was the legendary palace of Minos, the son of Zeus. Beneath the palace was, according to the myth, a labyrinth into which 12 Athenian youths, men and women, were placed only to be devoured by the terrible bull Minotaur—the Bull of Minos. Evans had no idea what language exactly the Minoans spoke except that he was positive that it was not a known language such as Greek. He was so certain of this, and such was his prestige, that anyone who dared to disagree with him lost not only any chance of working at Knossos but were removed from the British School and thereby lost any chance of doing archaeology in Greece.

The 18-year old Ventris did agree with Evans and proposed that the underlying language was Etruscan, arguing that the Etruscans had emigrated from the Aegean to Italy and might have left their language on an Aegean island such as Crete. Etruscan was, and is, a language that has never been fully deciphered, so it was a possible candidate. But Ventris could not properly fit the characteristics of the Etruscan language onto Linear B, although he kept it as an option until he actually

---

[72]For a biography see, The Man Who Deciphered Linear B: The Story of Michael Ventris by Andrew Robinson, Thames & Hudson, London, 2002. See also Ventris's collaborator John Chadwick's wonderful book *The Decipherment of Linear B*, Second edition, Cambridge University Press, 1967. This edition was reprinted in 1992 with an invaluable new postscript. Chadwick, who taught at Cambridge, died in 1998 at the age of 78.

succeeded in deciphering the language in 1952. (Etruscan uses the Greek-Phoenician alphabet and is usually written from right to left, unlike Linear B. To this day it has not been fully deciphered.)

It is this deciphering and its significance that will be our primary concern, but first I want to explain how I became interested in it. Eventually we will end up back at the National Archaeological Museum in Athens.

A few years ago I began research on a biography—*Dawning of the Raj*[73]—that was about the first governor-general of India, Warren Hastings. Hastings served from 1774 to 1784, after which he faced an impeachment trial which lasted seven years before he was acquitted. In doing this research I quickly became aware of a potential hazard. The supporting cast of characters was so interesting that unless one exercised considerable discipline one might easily write an encyclopedia rather than a manageable book. One of the figures who caught my attention was William Jones, the subject of the previous essay. The last language he learned was Sanskrit. This was of particular interest to Hastings who, as I discussed, had financed one of his protégés, Charles Wilkins, to travel to Benares to study Sanskrit with the local pandits. But no non-Indian had ever learned Sanskrit with the depth that Jones did. He introduced, by his published translations of Sanskrit literature—such as the play *Sakuntala* by the "Shakespeare of India", Kalidasa—this writing to Europe. He referred to Sanskrit as his "best" language.

During the course of learning Sanskrit, Jones confirmed what other European students of the language had already noted. There were remarkable parallels between Sanskrit and many of the European languages such as Greek. We will give a few examples, but first, let's look at a few of the words. The word for 'brother' in Sanskrit is *bhratar*, while in Greek it is *phrater*. The rules governing these sound changes form a whole discipline in itself. A second example involves both a sound change and something else, something that illustrates how language mirrors history. The word for 'door' in Sanskrit is '*dvar*' or '*dwa*r' depending on how you like your transliterations.[74] If you look in a modern Greek dictionary you will find '*porta*' given as the word for 'door,' which looks bad. But '*porta*' is a Latin loan word that reflects the Roman occupation of Greece that began in 200 B.C. However, if you look in a more complete Greek dictionary you will find a second, Old Greek, word for 'door,' '*theera*,' which is closer to the Sanskrit and illustrates another sound change rule. (The fact that '*door*' and '*dvar*' resemble each other should also be noted.) Numbers are a valuable clue. 'Two' in Sanskrit is "*dva*," while in Greek it is "*duo*" and so on up and down the number scale. But, there are also grammatical parallels. To cite one that is simple to explain, both Greek and Sanskrit have a special form for two of anything as opposed to, say, one, or three, or more. Both Greek and Sanskrit have verbs in which the first person singular ends in "mi," the second

---

[73] IVan R. Dee, 2000, Chicago.

[74] I am grateful to Professor Rosane Rocher, who taught Sanskrit at the University of Pennsylvania, for all her help.

person in "si," and the third person in "ti." Clearly something is going on here that goes beyond mere coincidence.

The other people who had noted these parallels usually offered the explanation that India had been invaded by Alexander the Great in the third century B.C., and that the Greeks had implanted colonies, traces of which can still be seen in places such as Taxila, which is in northern Pakistan. They argued that the Greeks had imposed their language on the indigenous population so that Sanskrit was actually some sort of Greek dialect. This explanation, which is superficially plausible, did not satisfy Jones. Moreover he had a platform from which he could propose his own explanation. As noted in the last essay, soon after he arrived in Calcutta, with Hastings's blessing he created the Asiatic Society, which he modeled after the Royal Society in Britain. The Asiatic Society was dedicated to all manner of things having to do with the Indian subcontinent. There were discussions of flora and fauna, of discoveries having to do with Indian history, of mathematics and astronomy, and of one of Jones's special interests, chess. Beginning in 1788, Jones created a journal, *Asiatic Researches*, which he largely wrote and edited. It was eventually circulated among interested scholars in Europe. Jones gave an annual address to the Society—there were eleven in all—in which he described what had interested him over the previous year. These were published in *Asiatic Researches*. In 1786, Jones delivered his Third Anniversary Discourse. It contains a paragraph, almost casually placed, which helped to found the discipline of historical linguistics. You will find this paragraph quoted in every textbook on the subject. It reads:

The *Sanscrit* language, whatever be its antiquity, is a wonderful structure—more perfect than the *Greek*, more copious than the *Latin*, and more exquisitely refined than either. Yet it bears to both of them a stronger affinity, both in the roots of verbs, and in the forms of grammar, than could possibly have been produced by accident; so strong indeed, that no philologer could examine them all three, without believing them to have sprung from the same common source, which, perhaps, no longer exists. There is a similar reason, though not quite so forcible, for supposing that both the *Gothick*[75] and the *Celtick*, though blended with a very different idiom, had the same origin with the *Sanscrit*, and the old *Persian* might be added to the list.[76]

Jones did not say specifically what he meant by Sanskrit's having a more "copious" structure than Greek, but here is an example of what he must have had in mind.

---

[75] An example can be found in J. P. Mallory's book, *In Search of the Indo-Europeans,* Thames & Hudson, London, 1989, p. 16. The word for 'field'

Sanskrit	Ajras
Greek	Agros
Latin	Ager
Gothic	Akrs

I would like to thank Professors Mallory and G. C. Horrocks for helpful correspondence.

[76] The full Discourse and much other material can be found in Sir William Jones: A Reader. Edited with Introduction and Notes by Satya S. Pachori, Foreword by Rosane Rocher and a Preface by Peter H. Salus, Oxford University Press, Delhi, 1993.

In Greek there are five grammatical cases, while in Sanskrit there are eight. Jones must have found it very difficult to imagine a collision of two languages, of the kind that his predecessors proposed, in which the language that is being replaced—in this case Sanskrit—ends up being more grammatically complex than the language that is supposedly replacing it, in this case Greek. Jones was sure that it hadn't happened and that, in fact, both Sanskrit and Greek had derived from the same proto-language, which might not even exist now.

While modern scholars can find things to criticize in this statement of Jones's, most of them acknowledge that it transformed the study of linguistics. Jones himself did not live to see this happen.

But what was this proto-language?

Early in the nineteenth century it acquired a name: Proto-Indo-European, or PIE as all the textbooks call it. But what was it? Who spoke it? Where? When? How did it become Sanskrit and Greek and the vast array of other languages—the Indo-European "family" —now spoken over much of the world?

The simple answer is that no one is entirely sure. The literature on this, accumulated over some two centuries, is enormous and often contradictory. I regard all of this from the perspective of an outsider—a non-combatant—and give you what appears to be a majority view, fully aware that I may be treading on a mine field. Don't worry. Linear B is still part of this story, indeed a very important part.

Since Jones first postulated his theory, the Proto-Indo-European people—the PIE people—have been located as far north as the North Pole and as far east as the Pacific. Some racists saw them as blonde Aryans, a kind of master race. Most scholars now believe they inhabited what is called the Pontic-Caspian region. This is a region that stretches across Turkey, Georgia, Armenia, and northern Iran. The archaeological evidence for this is sketchy, especially since no written records of PIE have been found. It is not clear that this language even had a written counterpart. If this was where the PIE people lived—probably from the fifth into the third millennium B.C.—then such archaeological discoveries as have been made shed some light on them. Gravesites have shown some characteristic features, including the burial of what were probably important males with a horse, presumably sacrificed for the occasion. The horse was evidently important to the PIE people. This seems to be connected to the invention of the light chariot with its spoked wheels, which can be dated back to the third millennium. Prior to this there were wheeled carts with solid, very heavy wheels, pulled by oxen, a vehicle that could not go much faster than about 2 miles an hour. A light chariot pulled by two horses could certainly do 10 miles an hour, which increased the mobility of the population.

Since there are no written records of PIE, reconstructing it is an art as much as it is a science. To give an example,[77] the third singular present of the verb 'to be'

---

[77] This example is given, along with many others, in *Historical Linguistics, An Introduction,* 2nd Edition by Winfred P. Lehmann, Holt, Rinehart & Winston, 1973, New York.

is *asti* in Sanskrit, *esti* in Lithuanian, *esti* in Greek, and *est* in Latin. The PIE reconstruction proposes that it was *esti* in PIE. The notion is that the PIE *e* developed into the Sanskrit *a*, which is supported by other examples. But how did this linguistic invasion come about? How, and when, did the PIE people come to India and Greece, and why did they make these migrations?

It is difficult to find unanimity among scholars on any of these questions. Was it some sort of natural calamity? Earthquake? Drought? Or was it a desire for conquest that caused these migrations? If the latter, then it might be related to the invention of chariot warfare, which gave the two-man chariots with a driver and an archer a tremendous advantage over foot soldiers.[78]

Whatever the explanation, some time in the second millennium the PIE people began to migrate south to India and west into Europe, including Greece. What interests us are the effects on language. Let's begin with India. The situation here is very interesting because the Indian languages can teach us where the migration stopped. At the present time there are 18 Indian languages recognized by the constitution, the kind you will find on bank notes or postage stamps. But there are some 1,600 distinct minor languages and dialects. These are divided into two major groups. The languages of the southern half of India, such as Tamil, are known as Dravidian languages. These are not Indo-European languages. You will recall the Sanskrit for 'two'-*dva*. Well, the Tamil for 'two' is *irandu*. Clearly we are in some other linguistic space. Languages of the northern half of India are known as Indo Aryan languages. Languages such as Urdu, the national language of Pakistan, are also spoken in India, Bengali, Punjabi, Hindi, and so on. These languages are derived from Sanskrit. Urdu has many Persian loan words and an Arabic script, while the other languages have fewer Persian loan words and their own scripts. But these mixtures are all derivatives of PIE and have vocabularies that are closely related. It is as if the PIE people came into India in migrations that for one reason or another stopped in the middle of the country.

What about Greece? The consensus seems to be that the PIE people moved into what is now mainland Greece around 1600 B.C. Their language gradually dominated, transforming into early Greek. However, remnants of the original language remain, for example in place names such as Knossos and Athens, which are not Indo-European. But did they get to Crete and the other islands? Here is where Linear A and B once again enter our story. You will recall that the tablets—Linear B in this instance—that are in the National Museum in Athens looked as if they had been in a fire. In fact they had. All these tablets, wherever they were found, *had* been in a fire. Otherwise they would not have been preserved. The clay would have soon crumbled into fragments. But fire turned them into ceramics. There are two fires that seemed to have played this role in Knossos. The first was around 1450 B.C., which

---

[78] This point of view is adumbrated in Robert Drews' book *The Coming of the Greeks*, Princeton University Press, Princeton, 1989. I would like to thank Professor Drews for his friendly help.

produced the ceramic Linear A tablets, while the second was about 1250 B.C., which produced the Linear B tablets.[79] Archaeologists have some general confidence in these dates because of the fortunate geographical accident that Crete is only some 200 miles to the north of Egypt. For centuries there had been commerce between Egypt and Crete and Egyptian artifacts, such as statuettes and vases, were found buried in the proximity of the tablets. But these Egyptian relics can be dated, at least to a good approximation, since different styles are associated with the regimes of different pharaohs. This then raises the following point. If Evans was right, and Linear A and Linear B both represent Minoan, then the PIE people—usually referred to in the Grecian context as Mycenaeans—never successfully colonized Crete during this period. But was he right? Enter once more, Ventris.[80]

Early in 1950 Ventris, now working full-time as an architect, devised a questionnaire that he sent to 12 scholars whom he knew to be working in the field of Minoan languages. He wanted to know where they thought the field stood a half-century after Evans's discovery of the Linear B tablets. Had they made progress deciphering of the tablets? Ten of the people he sent the questionnaire to responded in some detail. Two refused. One of them, Alice Kober, a classicist from Brooklyn College, replied tersely that such questionnaires were a waste of her time. She apparently did not add that she was terminally ill with lung cancer, which would soon kill her at the age of 43.

As it happened, Miss Kober was one of the scholars whose work was crucial in the actual deciphering. If her life had been spared, she might have cracked the Linear B code herself. We will come back to what she did after we tell you about another discovery, this one observational. It has a rather romantic history.

One of Schliemann's great accomplishments was to lead to the discovery that Homer's Troy was a real place, which one can identify with the city of Hissarlik in what is now Turkey. Schliemann did not realize that there were nine different levels of Trojan settlements built over each other so he misidentified his discovery, which was an earlier city than Homer's Troy. The American archaeologist Carl W. Blegen, from the University of Cincinnati, led expeditions there from 1932 to 1938 that finally seem to have identified the real Troy.

But there was another Homeric legend that intrigued Blegen, the palace of Nestor, one of the Homeric kings. Where was it? Homer said that it was located in Pylos. But where was Pylos? The legend said, "There is a Pylos before a Pylos and there is another beside," which was not much help. Blegen decided to explore the

[79] This timeline is controversial. There is a school of scholars that maintain that Knossos was destroyed by fire no later than 1375 B.C. This is hard to reconcile with the apparent vigor of Cretan culture up to the twelfth century. Linear B was still used on ceramic vessels up to the twelfth century.

[80] There is a very rich literature on the deciphering of Linear B. Especially helpful books are *The Code Book*, by Simon Singh, Anchor Books, New York, 1999.

The Bull of Minos by Leonard Cottrell, Efstathiadis, Athens, 1999.

I have already cited John Chadwick's *The Decipherment of Linear B*, but there is also his book, *Reading the Past, Linear B and Related Scripts*, University of California/British Museum, Los Angeles, 1997. There are several websites where you can find "ponys" of Ventris's deciphering. I had one of them on my visit to the National Archeological Museum, and there is Robinson's book previously cited.

area around the modern town of Pylos, which is south of Mycenae on the coast, on the bay of Navarino. By some sort of divine inspiration, Blegen picked what he thought was a likely site, and his expedition began to dig. Their first trial-trench produced, miraculously, a large number of Linear B tablets and, by the end of that season, they had found some 600 of them. The case in the National Archeological Museum, referred to in the beginning of this section, has a sampling. The ones that were dug up in 1939 survived the war by being kept in a vault in the Bank of Athens. Sometimes it is better not to glitter like gold.

Blegen did not get back to Pylos until 1952, by which time Ventris had, as Blegen was to learn, deciphered the tablets. However, the 1939 discovery raised an obvious question. If Evans was right, how did this Minoan language come to Pylos, where the spoken language was not Minoan but Indo-European. One ingenious suggestion was that Minoan scribes had been imported from Crete and had made their language the language of record. It is certainly not unknown for an official language to differ from the spoken vernacular. For example, in Hastings's time, the diplomatic language of a good deal of India was Persian, reflecting the Mogul conquests. If one did not understand this, and one found Persian documents in, say, a Bengali or English-speaking population, one might be very puzzled. However, there were people who did not accept the Cretan import theory. One of them was the distinguished British archeologist A. J. B. Wace, who was banned from working in Greece by Evans when he let his doubts be known publicly. There is a nice piece of irony here in that Wace, in 1952, found Linear B tablets in Mycenae, outside the palace that Schliemann had excavated. Freeman Dyson told me that Professor Wace was a visitor to the Institute for Advanced Study in Princeton in the early 1950s and that by that time he could write in Linear B as fast as he could write Greek.[81] Incidentally, Linear B tablets have been found in small quantities in Tiryns, which is close to Mycenae and in Thebes, which is north of Athens. These tablets are actuarial, like the others, which suggests that Linear B was a language of record and not a Lingua Franca. There are also jars inscribed with Linear B. No complete Linear A tablets have been discovered anywhere but on Crete, although some samples of the script have been found throughout the Aegean on pottery and fragments of tablets. Now back to Miss Kober.

There are a couple of general points that need to be made here. The problem of Linear B was at the extreme end of the spectrum of difficulties for a decoder. The underlying language was unknown, and there were no multi-lingual tablets that might act as a "pony." (Actually there was one, but its existence was not revealed until 1952, after Ventris had finished his decipherment.) For example, the so-called Rosetta stone, which had been found in Egypt in 1799, was tri-lingual and one of the languages was Greek, so the stone could be deciphered, and that led in turn to the deciphering of Egyptian hieroglyphics. The code breakers at Bletchley, who cracked the Enigma, knew that the underlying language was German because the German

---

[81] The actual discovery of the first tablet at Pylos was made by one of Wace's young assistants, Margaret Dow. She later married Murray Gell-Mann, the inventor of quarks. Gell-Mann told me that when his late wife brought Wace the tablet he was engaged in a conversation with some Greek dignitary. It took some time for her to get his attention.

military was using it for their communications. The second point is that it was not known what the individual signs in Linear B stood for. There were three possibilities: they could represent an alphabet, they could be an ideographic system like Chinese, or they could be a syllabary with each sign representing a syllable—or they could be a mixture of the three. As early as 1900, Evans had suggested a syllabary, although he never identified any of the syllables correctly. In Linear B there are, as we have seen, ideograms, such as the head of a horse, but most of the signs look quite abstract. There are about 89 different ones. This tells us that it is unlikely that they represent either a pure ideogramatic language or an alphabet. The reason is in the numbers. To have a purely ideogramatic language, like Chinese, requires a huge number of signs. The standard Chinese vocabulary has about 5,000, but good dictionaries list as many as 50,000. Eighty-nine signs are too few for an ideogramatic language. On the other hand, they are too many for an alphabet. Russian, at the top of the scale, has 40 "letters" in its alphabet, while Greek has a mere 24. The Greek alphabet, by the way, was a Phoenician import that came to Greece about 800 B.C. It is curious that Greek, which is Indo-European, should have an alphabet derived from a non-Indo-European, Semitic, language with a right to left written form. The Phoenician alphabet has 22 letters, all consonants. The Greeks added the vowels. Be that as it may, we shall see that a syllabic representation of a language poses problems that had to be faced in the decoding of Linear B.

The Linear B scribes had done something when they wrote their messages on the tablets that immensely aided the deciphering of the language. They had separated words-that are from two to eight symbols by putting a small, vertical line in between them, a kind of hyphen. If they had not done this, one wonders if it would have been possible to decode the language. Separating words was certainly crucial to Kober's work. It is at this point that one wishes for a word-processor that reproduced Linear B. Describing these signs accurately is extremely difficult. See the table below for the full set.

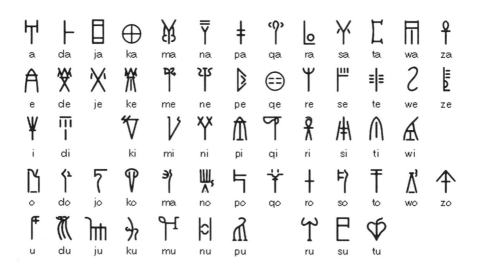

It is easy to imagine conveying the appearance of the English capital letters to someone who had never seen them over the telephone; an 'O' is a circle, a 'Q' is a circle with a tail, and so on. But what about Linear B? There is one sign that looks roughly like a snowboard, another that looks like two crossed hockey sticks, a third that looks like two parallel lightning bolts, and so on. The early decipherers, like Evans, tried to read meanings into these symbols that they didn't have. There was also the fact that each scribe had his own method of writing these signs so that they look somewhat different on different tablets. It is amazing that people like Kober and Ventris could keep them straight. Ventris seems to have had a remarkable visual memory. What most authors on this subject do is to replace each sign by a number, using a classification devised by the American archaeologist E. L. Bennett, Jr., just about the time that Ventris began his work. Bennett painstakingly went through the tablets, with their variant depictions of the signs, and reduced them to a canonical form to which a number could be attached. For example, the "snowboard" is number "37," the crossed hockey sticks "46,"and so on. What then did Kober do?

Put simply, she showed that Linear B was a highly inflected language and gave some suggestion as to how these inflections work. Actually Evans had suggested something like this in 1935, but without the detail. Since the tablets are largely actuarial accounts, the words are mostly nouns, some adjectives, but very few verbs. This means that declensions of verbs are not at issue here. What is at issue is how, or if, nouns, for example, take different forms in the singular and plural, the masculine and feminine, and the like. Are they inflected? What Kober noticed was that there were trios of words—later called "Kober triplets"—that had the same symbols at their beginnings but variants at their endings, something we are familiar with in English— 'draw,' 'draws,' 'drawing'—being an example. In terms of the numerical equivalents, sample Kober triplets were 25-67-37-57, 25-67-37-36, and 25-67-05. It is clear from this that 25–67 must be the stem of a word with different case endings—for example 57 and 36—showing that the language is inflected. But what is one to make of 25-67-05? At first sight one might say that 05, which is a symbol that looks a bit like a Cross of Loraine, is simply another ending. What makes this unlikely is the existence of a second triplet that takes the form 70-52-41-57, 70-52-41-36, and 70-52-12. You notice that the first two forms have, as before, 57 and 36 as endings, but the third has 12 as an ending. This suggests that neither 05 nor 12, a symbol that looks something like a glider, is a case ending. What Kober proposed was that these were bridge syllables in which the first letter was a consonant that attached to the previous symbol, while the second letter was a vowel. *Fo* and *so* would be examples, and, in fact, this turned out later to be what the symbols 05 and 12 stood for. She made the general assumption that all the symbols that stood for syllables took the form of a consonant followed by a vowel.

You will notice in this example how important it was to have a sufficiently large sample from which to work. If Kober had not found the second triplet she could not have reached her conclusion. Likewise, it required a large enough sample for her to have made her other important observation. As we have noted, the tablets appear to be for the most part actuarial documents. There are nouns followed by numbers, which must refer to the number of objects described by the nouns. Some of the nouns, such as 'man,' 'woman,' 'girl,' and 'boy' are represented by ideograms. But there are totals for, say, men and boys in which the numbers have been added up, including fractions. Part of the number system involves fractions. Next to the total there is a word that must stand for 'total.' However, this word is gender specific—a different word for the total of men and boys as compared to women and girls—which begins to disclose how gender is treated in the language. This, of course, is a far cry from deciphering the tablets, or even deciding what the underlying language was. Whether Kober, had she lived, would have taken the next steps we do not know. But apart from the specific importance of her observations, there was the matter of method. The previous decoders, like Evans, had taken a somewhat romantic view of the job. They liked to read into the symbols all sorts of arcane meanings, which produced some remarkable, and incorrect, results. Kober was the first person to look at the problem in a clear-headed, logical manner, and she got the furthest-until Ventris.

Once the results of his 1950 survey were in, Ventris put them together and sent copies to the participants. He ended the text with the statement that he was withdrawing from the field since there were plenty of people to carry on. He was now working as a full-time architect for the Ministry of Education and had a wife and family to support. But Ventris could as easily stop thinking about Linear B as he could turn off the force of gravity. Furthermore, in 1951, there was a very important development: Bennett published the transcripts of the Pylos tablets that Blegen had found in 1939 and which he had had photographed so that his student Bennett could study them. These were the ones—some 600—that had survived the war in the Bank of Athens. The reason that this was so important was the way Evans had treated the tablets he had found in Knossos. For whatever reason, he dribbled them out so that when Ventris began his work only 142 out of 2,846 tablets, and fragments of Evans's tablets, had been transcribed and published. It turned out later that he had missed many of the fragments that archaeologists were able to restore. What Evans would have thought of any of this activity, to say nothing of what Ventris finally did, will remain unknown. He died in 1941, at the age of 90. The beautiful house he had built for himself near Knossos was, after the fall of Crete, used as the headquarters of the German general in charge of the occupation. In any event, Ventris now had enough material to begin working on the decoding.

It is unclear whether Ventris had had any formal training, in say, the air force as a cryptographer—a code breaker—or if he simply invented the techniques for himself. What cryptographers do when confronting an unknown code is to look for patterns, symbols that are used more often and with higher statistical frequency than others. If there is a language involved that, unlike Phoenician, expresses vowels in its written form, then these vowels may stand out because of their frequency of use and their placement. Of course, someone constructing such a code will take pains to obscure this if the code is meant to be secret. But the Linear B scribes were certainly not trying to devise a secret code. They were simply keeping records. Therefore there was no reason for them to try to hide these frequencies. Here is an example of how Ventris used this sort of frequency analysis. He had decided that the assumption that the symbols that represented a consonant followed by a vowel could not be the whole story. Suppose you wanted to use this system to represent the English word "firm." The "fi" is fine, but what about the "rm"? You could deal with this by introducing silent vowels so that you could render "firm" as say, *fi-ri-me*. This, of course, introduces a considerable clumsiness into the writing. But what do you do with "infirm"? The "nf" can be dealt with by introducing a silent vowel, but what about the initial "i"? Inserting a silent consonant in front of the "i" opens up Pandora's box.

The economic way of doing this representation is to have symbols that stand for pure vowels. This is what Ventris thought must have happened with Linear B. At the time, he had no idea how many different vowels there might be, but he did have a way of testing his pure vowel theory—the frequencies. He noted that certain symbols appeared with anomalously high frequencies at the beginning of words, for example 08, which looks like a double ax—and had been so interpreted by Evans, leading to all sorts of fanciful speculations. Ventris proposed that this symbol was, in fact, the pure vowel "a." Eventually, he identified five. This was a tremendous conceptual breakthrough. There were others.

Ventris noted that the symbol 78, which is a circle with four hyphens in the middle arranged in two parallel rows, occurred very frequently at the end of words. Ventris conjectured that when this symbol, which also occurred elsewhere in words but much less often, was used at the end of a word, it was probably a conjunction, most likely "and." This turned out to be right. He also vastly expanded Kober's inflections. By August of 1951, he had found 159 words in the Pylos tablets that appeared to show inflections. In particular he found sets of gender specific words that had varied endings. He used these in a very ingenious way, taking advantage of his discovery of the pure vowels. He reasoned that the symbols for these endings would have the same consonants but different vowels, a masculine and feminine ending, such as señor and señora. Thus, he began building up what he called a "syllabic grid." This was a diagram that was divided into rows and columns. Each column was headed by a vowel and each row represented a different consonant. He still did not know what all these vowels and consonants actually were, but, with this arrangement, once he determined some of them, there was every chance that the language would begin to collapse like a house of cards.

While he was doing this work, Ventris sent out progress reports. They are fascinating. They look like the lab reports of a very meticulous scientist. In addition to his grids he made comments as he went along. By November of 1951, he began to have an intimation that the language he was dealing with, which he still thought was Minoan, might have some Greek influences. He wrote, "The latter [Greek forms] are also worth considering on the remoter possibility that the Knossos and Pylos tablets are actually written in Greek, though I feel that what we have seen so far of Minoan forms makes this unlikely."[82] By March of 1952, he was back trying

---

[82] See Chadwick, *The Decipherment of Linear B*, op. cit., Chapter 4, for a discussion of these reports. This quote is found on p. 60.

to fit his syllabary to Etruscan. Etruscan is an interesting case because the inscriptions are perfectly legible, thanks to the fact that both Etruscan-Latin and Etruscan-Phoenician bilingual examples have been found. It is just that no one has been able to figure out what the words mean. Ventris knew enough to realize that his increasingly precise Linear B grids were not Etruscan. Then came the decisive step.

In re-examining Kober's "triplets" and examples like them, Ventris made the remarkable conjecture that some of these could be place names such as towns. But which and where? Crete seemed like the most probable location. To take the next step took he advantage of studies of the Cypriot written language. He had been aware of these, but up to this point, they had led nowhere. This written language is also syllabic, and several of the symbols look very much like Linear B. Indeed, they were very likely developed from Linear A. Because of some bi-lingual Cypriot-Phoenician tablets, and some that used both the local language and the Greek alphabet, this syllabary had been decoded in the nineteenth century. But identifying the two syllabaries too closely had led to these blind alleys. The symbol for *pa*, for example, in the two syllabaries are the same, but the vowels in the two languages are represented by totally different symbols. On the other hand, some symbols that look identical actually correspond to very different syllables. However, Ventris decided to use the more certain identifications to try to deconstruct at least some of the Linear B symbols, and then to test this by finding a town in Crete that would fit. He had already concluded that the double ax symbol-8, because of its frequency at the beginning of words, was probably "a." He identified 06, which looks like a thatched roof on top of a column, with the Cypriot "na." Thus, using his grid, he was able to identify one of the consonants as "n" By similar sorts of arguments he identified the vowel "i." Again using the grid he was able to guess at the symbol for "ni", a symbol that looks vaguely like a flower. Thus armed, he began looking for names of ancient towns on Crete that might fit. Homer had mentioned a Cretan town which he had called *Amnisos*. So Ventris looked through his tablets to see if there were any words of the form 08…30. He found four. The shortest one was the word 08-73-30-12. From his grid he saw that 30 and 73 had the same vowel 'i.' Thus the first three symbols must spell out *amini*. The vowel 'i' between the 'm' and 'n' is inserted as a silent vowel to complete the syllable. The number 12—a symbol that vaguely looks like an arrow in flight—must, he reasoned, stand for "*so*." The final "s" had been dropped. The three longer words were then inflections whose grammatical function he could now begin to determine. By applying the same sort of reasoning he was able to decode *Tulissos*, a significant town in central Crete, and *Knossos*. While this was quite marvelous, it did not shed any light on what the underlying language was. The names for these ancient towns presumably reflected whatever language was spoken when they were founded, which could be quite different from Linear B. Just because a town in upstate New York is called "Oswego" does not mean that its inhabitants now speak Iroquois.

Ventris next turned to a commodity that was mentioned on both the Pylos and Knossos tablets. Here he came up with something like the Greek words *korannon* or *koliandron*, both of which refer to the spice 'coriander.' This was interesting, but it didn't really prove anything, since these could have been loan-words from Minoan.

At this point Ventris focused on the declensions of the words for 'boy' and 'girl.' He found candidates that seemed Greek, but not any Greek that he was familiar with. The same thing occurred when he looked at declensions of "total." He also found that the Knossos tablets with chariot ideograms had a word that looked like the Greek word for "chariot." "The Greek chimera," he wrote, "again raises its head."[83] He simply could not persuade himself as yet that, in fact, Linear B *was* Greek, but a Greek so archaic that Homer or Plato might have had great difficulty in understanding it. It as if a Greek archaeologist had dug up tablets that looked as if they might be the English spoken by Shakespeare, but were in fact written in the language of Beowulf.

By June of 1952, Ventris was pretty sure, but not yet certain, that Linear B was a form of written Greek. As it happened, at this time the second volume of *Scripta Minoa*, which contained more of the tablets that Evans had found at Knossos, a century and a half earlier, had just been published. The B.B.C. Third Programme had chosen to present this event to a presumably minuscule audience. As far as the B.B.C or anyone else knew the deciphering of Linear B was still an unsolved problem with rather little progress towards its solution. Ventris was invited to speak. One wonders who suggested him since he had no public reputation as an expert on the subject. He was, after all, an architect. As always, Ventris was very candid about where things stood. He explained that he thought that Linear B was a form of Greek, but a form so far removed from any Greek that he knew that this posed problems that he could not solve. Among his listeners was a young linguist named John Chadwick. As it happened Chadwick, who specialized in ancient Greek dialects, had just taken a position at Cambridge. He had never been in the Evans camp and had even tried his hand at deciphering Linear B on the assumption that it was Greek, but without having enough of the tablets in hand to make progress. He consulted one of the senior professors at Cambridge who had been in touch with Ventris, and who had the copies of the latest grids. Chadwick examined them and put his knowledge to work deciphering several words that had stopped Ventris. When he wrote to Ventris explaining what he had found, he was immediately invited to become a collaborator. This certainly had to do not only with Chadwick's knowledge, but with Ventris's certainty that, now that the cat was out of the bag, he was going to be attacked from all sides. Very few of the professionals in this field were willing to entertain anything other than the Minoan solution, and they had a lot at stake. One of the exceptions was Professor Wace, who had thought it was Greek from the beginning. Expressing this had gotten him banned from practicing archaeology in Greece until after the war.

Professor Wace, though, lived until 1957, by which time Ventris's decoding was widely accepted. In the meantime, Chadwick could share the brunt of the inevitable onslaught. From Chadwick's description, and from other peoples' accounts, he and Ventris had a very happy collaboration. Not only did they bring complementary skills to the work but both of them were unassuming. There was apparently no ego

---

[83]Chadwick, op. cit., p. 66.

involved of the kind that destroys so many scientific collaborations. In fact, the more important the scientific discovery, the more fragile is the collaboration. These two men were sitting on one of the greatest discoveries in the history of classical studies, yet they went about their business in a cheerful and entirely amicable way. Pretty soon they were able to decipher whole sentences from the tablets. The first one was from a Pylos tablet. When deciphered it read, "At Pylos, slaves of the priestess on account of sacred gold: 14 women." While the meaning of this is obscure, it already tells us more about Mycenaean society than all of Schliemann's gold objects in the Archaeological Museum put together. What these objects tell us is that this society had skilled artisans, and that some people were wealthy enough to pay them to do this work. But what, for example, could we learn about the Russia of the Tsars from a case of Faberge Easter eggs? Would it tell us, for example, that much of this society was built on the labor of serfs? Compare this to "slaves of the priestess." A society that was three and a half millennia old was beginning to tell us about itself, in its own voice.

By the fall of 1952, Ventris and Chadwick were well underway with the writing of their great paper, "Evidence for Greek Dialect in the Mycanaean Archives." This title was very deliberately chosen. There was no equivocation that the Linear B tablets, wherever they were found, had been written by the descendants of the PIE people—that is, Greeks, not Minoans or anyone else. They had the paper done by November. It was accepted for publication in the *Journal of Hellenic Studies* and was scheduled to appear the following June, almost rocket speed for this kind of publication. However, both of them still had doubts. But something happened in May of 1953 that changed everything. Ventris received a letter from Carl Blegen. Blegen, it may be recalled, had found the first Linear B tablets in Pylos in 1939. These were the ones that had survived the war in the Bank of Athens. But in 1952 he was back in Pylos, where he had found more tablets. It had taken him some time to clean them; they had been underground for some 3,000 years. However, there was one he thought it was imperative that Ventris see. Blegen had had advanced word of the deciphering, and even had the code. Using it he had partially deciphered a tablet that he had logged in with the number P641. (See the first illustration above.)

This tablet, as it happens, is in the center of the display case in the Archeological Museum. It is unlikely that you would notice it except that a laconic white card in the case tells you to look for it. Even with the card, it is unlikely many people understand what the point is. The tablet is midsize, as Linear B tablets go. It looks as if it started out to be a rectangle whose sides got curved when it was made. All these tablets were meant to be held in one hand while the scribe wrote with the other. It is about 5 or 6 inches long and perhaps 3 inches wide.

When I leaned over the cabinet to try to measure the tablet more precisely through the glass, I was shooed away by the museum guards. It is a medium brown; light enough so that the writing is quite easy to read. There are three lines of script. Apart from the words and numbers, there are ideograms of vessels. They almost look like what a child would draw if you asked the child to draw something you could put water in. There are eight of these little caricatures. Two of them are shown standing on stands and the rest would presumably rest on the ground. On their tops, objects are drawn that look like ears, but must represent handles. There are no-eared, two-eared, three-eared, and four-eared ones.

I had come to the museum with a Linear B pony because I thought it would be fun to try to read some of the tablets. I had no idea in advance which tablets would be on display, and absolutely no inkling that this one would be there. I knew enough about this tablet in advance so that I could have guessed at the meaning of the words without the pony. It was nonetheless wonderful to spell them out. Why the fuss?

This tablet is often called the "tripode" tablet, and with good reason. Let us start with the two that are resting on stands. The stands have three legs. One of them has the representation for the number two after it, while the second the number one; there are two of one kind and one of the other. The variety, of which there are two, is described by a sentence that begins *ti-ri-po-de* while the other, of which there was only one, begins with *ti-ri-po*. By now you should have enough of a feeling for Linear B to see that in *ri* the dummy vowel "I" has been inserted. The two words should be "tripo" and "tripode." You do not have to be a Greek scholar to remember what a tripod is a three-footed stand—although you would have to be reminded that in compound words of this type "tri" means "three," which otherwise would be "tria" in the neuter or "treis" in the masculine or feminine. You will also recall that the final 's' was left off in the Linear B rendering of words like "Knossos." Thus "tripo" should be "tripos." But why is the "de" added to the second "tiripo"? Remember that Sanskrit and early Greek have in common a special form for two of anything. "Tripode" is the dual form of "tripo." The modern Greek word for one "tripod" is "tripodo." "Tripode" indicated that there were two tripods of the kind shown. It all fit. You will now have no trouble understanding Ventris's excitement. Here, at last, was a bilingual tablet, one language being the ideograms and the other Greek.

And then there were the "ears." My favorites were the four-eared vessels and their description. I was delighted to use my pony to translate the Linear B signs that spelled out *qe-to-ro-we* that are written just before the ideograms. The "we" is a reference to "handle" and occurs in all these examples. It is related to the Greek word for "eared," "otoeis." What interested me was the *qe-to-ro*. The ancient Greek word for "four" is "tessera" in the neuter or "tessares" in the masculine or feminine. But here we have a word for four, "qetoro," that is even older. It is at least a thousand years older than the Greek of Plato. The Sanskrit word for four is "catvaras," "qetoro" — "catvaras" —the Greek of these Mycenaeans is merging with the Sanskrit, back towards their common PIE origin, the dawn of the Indo-European language family.

Most great theories have one particular experiment that establishes them beyond a reasonable doubt. For Galileo, it was dropping the balls of different weights from the Leaning Tower of Pisa and noting that they landed at essentially the same time, proving his theory of falling bodies. For Einstein, it was the solar eclipse expeditions of 1919, which showed that the Sun really does bend starlight as much as he said it did, and for Linear B it was the tripode tablet. Chadwick and Ventris's paper was published in June of 1953. On June 24, Ventris gave a lecture in London. As it happened, this was just after the first climb of Mount Everest had been reported, so inevitably the *London Times* referred to it as "the Everest of Greek archaeology."

It would take more than an essay like this to describe what has happened in this subject since Ventris's lecture. It would take volumes. The last time I looked, I found more than 86,000 websites devoted to Linear B. One offers to let you download Linear B fonts, and there are others with very nice ponies.[84] Most have to do with what has been learned about the Mycenaean world from the tablets and the related archaeological discoveries, making it clear that we are dealing with an extraordinarily complex and thriving society, but still one with slaves and possible human sacrifices. Here are two examples. The first is Linear A, which you will recall was the writing from which Linear B descended. There are only a small number of tablets that have so far been found, and most of these come from the Minoan palace of Haghia Triada in Crete. The characters often closely resemble Linear B characters, which is what one might expect. The assumption is made that the syllables in both languages correspond—which seems reasonable. Thus you can use Linear B as a pony for Linear A. What then is the problem? The problem is that when you do this with words such as "total," which appear to be used in the same way on both sets of tablets, you come up with the Linear A word-*ku-ro*, which does not seem to be in any language anyone can identify. So far, Minoan appears to be a language without a family. Perhaps more tablets will be found and the matter resolved.

The final example we will discuss involves what may be the deepest mystery in Mycenaean studies: Why did this civilization—the Late Bronze Age, as it is usually called because copper and tin were imported into Greece to make bronze—disappear? By 1200 B.C. the palaces had been burned to the ground, and Greece entered a dark age from which it did not recover for some four centuries. How did this happen?

The short answer is that no one knows for certain. However, the Pylos tablets offer some remarkable clues.[85] Most of the Pylos tablets are not dated at all. The inventories on them presumably refer to the current year, whatever that was, and the next year the scribes used either refurbished or new tablets. But these Pylos tablets have on them what looks like a month. The translation does not correspond to a name that is known, but from the tablets one can infer that it was probably spring. There is no clue as to the year. The tablets describe preparations that were being made for the defense of Pylos from the sea. Its situation was such that a land attack was so unlikely that the palace was not even fortified. The tablets speak about how to organize the relatively small number of defenders—some 800—to guard a coastline of about 150 kilometers. The tone is urgent. About the same time incursions were occurring in Egypt by a force that was described as "Peoples from the Sea." Who these were exactly is not known. What is known is that a force of invaders,

---

[84]If you want to look up these sites you will find that simply using "Linear B" in your search engine will turn up all sorts of sites devoted to linear algebra and the like. Something like "Linear B" + "Minoan" will restrict things.

[85]For one view and a discussion of other possibilities see, for example, *The End of the Bronze Age* by Robert Drews, Princeton University Press, Princeton. For a good discussion of the Pylos tablets see *The Mycenaean World*, by John Chadwick, Cambridge University Press, Cambridge, 1976.

perhaps the same, stormed the Palace of Nestor in Pylos and burned it to the ground. They may well have done the same at about the same time for Knossos, and for every other site where Linear B tablets have been found. A civilization was destroyed, but some of its records were preserved.

What of Ventris? He became a sort of public figure, giving lectures, receiving awards, and writing articles to explain the discovery. Whether he would have remained an architect or drifted into academia we will never know. On September 6, 1956, while driving home late at night, his automobile collided with a lorry. He was killed instantly. He was 34.

Chapter 11
In a Word, "Lions"

> *The Etruscans, as everyone knows, were the people who occu-*
> *pied the middle of Italy in early Roman days and whom the*
> *Romans, in their usual neighborly fashion, wiped out entirely*
> *to make room for Rome with a very big R. They couldn't have*
> *wiped them all out, there were too many of them. But they did*
> *wipe out the Etruscan existence as a nation and a people.*
> *However, this seems to be the inevitable result of expansion*
> *with a big E, which is the sole raison d'être for a people like*
> *the Romans.*
>
> —D. H. Lawrence[86]

For reasons that will become clear later, I have been perusing Etruscan glossaries. The one I have in front of me is typical of the genre. It is given as an appendix to *The Etruscan Language; An Introduction*[87] by Giuliano Bonfante and his daughter Larissa Bonfante. The senior Bonfante, who died in 2005 at the age of 100, was a professor of linguistics at the University of Turin, while Larissa Bonfante is a professor of classics at New York University. Both are well known in the field of Etruscan studies. The glossary is not very long, a few hundred words, many of which have question marks after them, indicating that the proposed meanings are tentative. Most of the words look very strange. That is, they do not seem to correspond to any language one knows. For example "fleres" apparently means "statue, " while "nurthanatur" is a group that does "nurth"—whatever that is. The glossary gives a question mark after "nurth. "

As I was looking down the columns I came across a word that stopped me in its tracks— "leu," which means "lion." Leu the lion, where did that come from? To put this in perspective, one must understand that Etruscan is in a certain sense an orphan

---

[86] *Etruscan Places* by D. H. Lawrence, Nuova Immagine, Siena, 2001, p. 31. I have preserved Lawrence's odd diction.

[87] *The Etruscan Language; An Introduction*, 2nd edition, by Giuliano Bonfante and Larissa Bonfante, Manchester University Press, Manchester and New York, 2002. I am grateful to Professor L Bonfante for several e-mails and for sending me two issues of Etruscan News, a newsletter of the American Section of the Institute for Etruscan and Italic Studies. I have since subscribed.

language. Like Basque, it does not belong to any of the well-known linguistic families such as the Indo-European, the Semitic, or the African language families. There are scholars who claim to see similarities between it and the Raetic language, which is found on some inscriptions in northern Italy or Lemnian, a language found on artifacts from the island of Lemnos. Perhaps these three languages descended from a prehistoric proto-language or maybe the Etruscans got around. But "leu"?

The *American Heritage* dictionary is not much help. On the etymology it notes that our lion word comes from "Middle English, from Old French, from Latin *leo,leon*, and from Greek *leon*, of Semitic origin." It then goes on to say, "Old French *lion* is the source of English *lion*, and the Old French word comes from Latin *leo*, *leonis*. After that the etymology is less clear. The Latin word is related somehow to Greek, *leon*, *leontos* (earlier *lewon*, *lewontos*), which appears in the name of the Spartan king *Leonides*, "Lion's son," who perished at Thermopylae. The Greek word is somehow related to Coptic *labia*, *laboi*, "lioness." In turn, Coptic *labia* is borrowed from a Semitic source related to Hebrew *labi* and Arcadian *labbu*. There is also a native ancient Egyptian word, *rw* (where *r* can stand for either *r* or *l* and vowels were not indicated), which is surely related as well. Since lions were native to Africa, Asia, and Europe in ancient times (Aristotle tells us that there were lions in Macedon in his day), we have no way of ascertaining who borrowed which word from whom." This is all very well, but what has it got to do with Etruscan? Before we reveal that, first we need an historical and linguistic detour.

It is not certain when the Etruscans first came to Italy, nor from where, but by the eighth century B.C. they had amalgamated small settlements into what became the Etruscan cities in what is now Tuscany. They called themselves *Rassena* or *Rasna*, but were called various things related to the word "tower"— "*Tursci*," or people who build towers—by their neighbors, who had only lower structures.

Most of the cities in "Etruria" were close to what is now known as the Costa degli Etruschi, the Etruscan coast of the Tyrrhenian Sea. Many of the cities such as Ovieta and Bologna will be familiar, while a place such as Vetulonia may not be. I have a particular fondness for Vetulonia, since on a recent bicycle trip I visited it, which accounts for my recent burst of interest in things Etruscan. Since Vetulonia is on top of a very steep hill I must, in the interests of full disclosure, say that I did not peddle to the top but managed to cop a ride in a van. In Etruscan times it seems as if the town stood on the shores of a lake that communicated with the sea, so that it was a port. There are the remains of the original town and some tombs of the kind that contained much of the statuary, gold work, coins, and the like, which tell us the little that we know about the Etruscans. Some of this can be found in a charming small museum in the modern town. A much better collection, of course, can be found in the Museo Archeologico in Florence. Among other things it contains the magnificent Chimaera of Arezzo. This bronze figure—a specialty of Arezzo—and was made around 400 B.C. and apparently restored by Botticelli. It has the body of a lion with a snake for a tail. The snake is attacking the horn of a goat that is growing out of the lion's back. The lion figures large in Etruscan art, although there were surely none in Etruria at this time. There are also some examples of the engraved copper mirrors—some of the engravings are quite sexual—which occasionally have inscriptions on them that seem to identify the owner. One has the impression that the Etruscans, like all Italians, were fond of eating and drinking and sex.

The Etruscans apparently traded widely with their neighbors. One of their most important imports was the alphabet, which they got from the Greeks, who in turn had gotten it from the Phoenicians, who apparently invented it—and a magnificent invention it was. It meant that everything you could say in a language could be written down with a small number of symbols. Phoenician writing, like the other Semitic languages, was from right to left and did not express the vowels. The Greeks wrote from left to right and did express vowels. The Etruscans took this

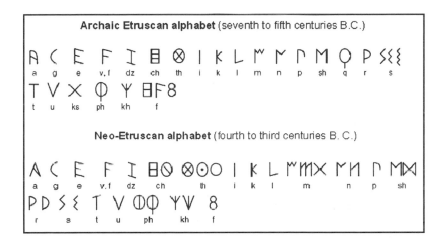

over with a few variants. But the symbols do not look like the modern Greek alphabet. Above is an illustration of what they did look like.[88]

The numbers are also interesting. I bought a T shirt in the museum in Vetulonia that had them. "C" stands for a hundred and "X" for ten, like Roman numerals, except that the Etruscans used this notation first.

One might think that knowing the alphabet one would have no trouble deciphering the language. Alas this is not so. As we have seen in the last essay, when it comes to language decoding there are three cases one can consider. There is the case of a language like the Egyptian hieroglyphics, where both the symbols and the underlying language they represent are unknown. Then there is the case of a language like Linear B, found on the tablets in Crete, which date back a few hundred years prior to Etruscan writing. We have seen that the symbols stood for syllables and that the underlying language turned out to be an archaic form of Greek.

---

[88] For a number of alphabets the reader can consult the website omniglot.com or the Bonfante book. This version is taken from the website.

In the Etruscan situation the alphabet is known, but not the underlying language. Moreover, while thousands of fragments of Etruscan writing have been found, they are not that helpful. Most of them are proper names belonging to people in tombs or on vases or mirrors. There are a few notable exceptions. One is a bronze model of a sheep's liver, which has on it inscribed the names of a large number of gods. One supposes that this was a teaching tool used to illustrate how to use sheep's livers for prophecies. Even when Roman power was increasing, Romans were sent to Etruscan cities to learn the art of divination. There is also a mummy wrapping that has a fairly extensive text involving religious ceremonies. But if the Etruscans had poets or historians, their work is still undiscovered. One is basically trying to decipher a language from what is written on tombstones. This brings us back to where we started, the lion.

It was the Egyptians who introduced the scarab, a gem in which the top is in the form of a beetle and the bottom is a carved surface that could be used as a seal. These scarabs found their way into Greece and then into Etruria. Whether the Etruscan scarabs were carved by Etruscan artists or by Greeks working in Etruria is impossible to say. This has some bearing on our lion. As mentioned earlier, the lion plays a very important role in Etruscan art, including scarabs. There are scarabs that show individual lions or lions attacking various forms of prey. But among them there is one that is quite unique.[89] It depicts a lioness and a cub in the act of suckling. Above the lioness there are three symbols. They are "l," "e," and "u" in the archaic Etruscan alphabet. This represents our entire knowledge of the Etruscan word for lion. It is what linguists call a hapax legomenon, a word or form that occurs only once in the recorded corpus of a language. It presumably was taken from the Greek—a loan word—but it is very strange. It is not the Greek word for lion, to say nothing of lioness, which is "leaina." What does this mean? Dieter Steinbauer is one of the acknowledged experts in the Etruscan language. Here is what he wrote: "There are three observations to be made. As Etruscan was a language that didn't normally distinguish between masculine and feminine genders [Male and female personal names did have markings in Etruscan], "leu" must have the meaning "lioness" too. The loss of the final 'n' is embarrassing because Etruscan words with an ending 'un' do occur. Perhaps the loan passed via an Italic language where n-stems had no 'n' in the nominative (e.g., Latin "homo"; Greek Platon-Plato). Normally, nouns were rendered in the accusative of the source language. So perhaps the Etruscans thought the animal 'animated.' I know no further example."

That is all that we know. But Etruscan studies are very active, so that one can hope that another "lion" will eventually be found.

---

[89] A reader who wants to see a photograph will find one in *Archaic Greek Gems* by John Boardman, Thames & Hudson, London, 1968. The scarab in question can be found on page XXXI of the photographic section at the end of the book. It is number 446.

Part IV
Fiction and Stranger than Fiction

Chapter 12
The Pianist, Fiction and Non-fiction

When Warsaw fell to the Germans on September 28, 1939, effectively ending Polish resistance, it is now estimated that there were about 3.5 million Jews living in Poland. When the war ended, some 200,000 were still alive. Among the survivors were film director Roman Polanski and composer Wladyslaw Szpilman. Polanski, now in his seventies, was a young boy, while Szpilman, who died in the year 2000 at the age of 88 in Warsaw, was in his thirties.

Like the rest of the Jewish survivors, both men had good luck, determination, and the help of others. If a Pole was caught hiding a Jew not only was he or she killed, but so was the entire family. Nonetheless, heroic Poles hid Jews, including Polanski and Szpilman. Polanski was from Krakow, while Szpilman, who was a pianist and composer, was from Warsaw. The two men first met in Los Angeles in 1967, and then, many years later, in Warsaw. At this time Polanski optioned Szpilman's account of his wartime experience for a film. The first version of Szpilman's memoir, entitled *Death of a City*, was written in Polish, and published in 1946. It was then not reprinted for over a half century. The new version, *The Pianist*, was published in English in 1999.[90] There were attempts to bring it out in Poland and Germany in the interim, but they were always blocked by censorship of one kind or another.

Seeing Polanski's film or reading the journal is a shattering experience, but they are not the same experience, and that is what we will discuss. Polanski has not produced a documentary, although what he puts on the screen is based on what actually happened. But much is left out, and, much of what is there is sometimes enhanced or distorted in the interests of telling a story. Polanski's film begins with the war and ends when it ends. We are not told what happened before or afterwards. This makes some of the film less comprehensible than it might have been. For example, we are not told that Szpilman had studied the piano with Artur Schnabel in Berlin, leaving Germany only when Hitler came to power in 1933. This explains two things that are puzzling in the film. There is a crucial moment—I will discuss it more fully

---

[90]*The Pianist*, by Wladyslaw Szpilman, translated by Anthea Bell with extracts from the Diary of Wilm Hosenfeld, foreword by Andrez Szpilman and epilogue by Wolf Biermann, Picador, New York, 1999.

J. Bernstein, *Physicists on Wall Street and Other Essays on Science and Society*,
© Springer Science+Business Media, LLC 2008

later—in which speaking German appears essential to Szpilman's survival. We are offered no explanation of how Szpilman acquired this skill. The second is something that both the book and the film have in common. There is no sense of hatred towards Germans. That is left for the viewer or reader. Szpilman never speaks of revenge. When he was asked about this he responded, "When I was a young man I studied music for two years in Berlin. I just can't make the Germans out…they were so extremely musical!"[91]

In his own strange way, Szpilman has put his finger on something very profound, something that is left out of the film. It is not clear how it could have been included without distorting its shape. In the film there are only two kinds of Germans— vicious, physically repulsive sadists and, in one case, the good looking army officer who saved Szpilman's life. The deeply disturbing thing about the Holocaust is that it was largely carried out by perfectly ordinary people. There is one particularly shattering photograph of an SS officer in the process of shooting Polish Jews who had been forced to dig their own graves. The scene was being witnessed by his fiancé, who had come for the entertainment. They looked like the sort of couple you might see taking a Sunday picnic in a park. If there is a lesson to be learned from the Holocaust it is that there is a residue of evil in most of us and, if this is manipulated, any one of us might become this officer. This is not to excuse the Germans for what they did. There is no excuse possible. But it is not a lesson that you will learn from Polanski's film.

The film and the book begin in different places. The film begins with Szpilman giving a Chopin recital on Polish National Radio on September 23, 1938—the last live concert on Polish radio until the end of the war. (The film ends with Szpilman returning to the same recital and then playing a concerto with an orchestra.) The recital is interrupted by the bombing of the city. The rest of the film proceeds chronologically. The book begins at the sealing of the Warsaw Ghetto in November of 1940. After that date any Jew caught trying to leave the ghetto was to be shot on sight. It then works backwards to the events that led up to this. The film can only suggest what life in the ghetto was like. In fact, for some time there were two ghettos. There was a "small" ghetto in which the more prosperous Jews, like the Szpilmans, lived and a "large" ghetto that had been packed with poverty-stricken Jews moved from wherever they had lived. There was only one passage between these ghettos, which the Germans opened and closed on various whims. One of the more memorable scenes in the film and the book was that of German policemen forcing arbitrarily created couples—cripples, for example—to dance to ever faster music until they dropped from exhaustion.

The film, although it makes a valiant attempt, never fully captures the life in the ghetto. In the beginning it was fairly normal. The wealthier Jews frequented restaurants and cafes. Indeed, Szpilman's book begins by describing how he supported his family by playing the piano in these cafes. This is also shown in the film. What is not conveyed is what the crowding of the ghetto meant in terms of

---

[91] *The Pianist*, op. cit., p. 213.

people's health. The ghetto was so crowded that when anyone went out on the street, say to shop, it was impossible not to collide with other people. This meant that lice, which were ubiquitous, passed from person to person constantly. These lice carried typhus from which many people died. Szpilman describes how his mother and father and his two sisters and brother constantly tried to pick lice off each other. He also notes that if one had enough money one could buy a dose of anti-typhus vaccine. He had enough for one dose, but decided not to buy it for himself since he could not afford to buy the vaccine for his entire family. Considering what was going to be their fate, no amount of vaccine would have been enough. Polanski's film does not fully depict the squalor, which got worse and worse as the ghetto became more crowded, in which people lived. No film can convey the stench, much of it from dead bodies.

Until the summer of 1942, it was not clear to Szpilman, or to anyone else, exactly what the Germans had in mind. There were raids in which people were taken off and shot, but these appeared to be haphazard events involving relatively small numbers. But then there was a change. The Germans began a systematic attempt to liquidate the ghetto completely. By this time the extermination camps at places like Auschwitz and Treblinka were operational. They were first tested out on Poles picked at random off the street. As mentioned in the previous essay, a Polish colleague of mine, Jacques Prentki, was picked up in such a raffle. He escaped by jumping from the box car of the slowly moving train taking him to one of these camps. He told me that when he got back to Warsaw he was ready to kill the first German he saw.

In the summer of 1942, the wholesale deportation of ghetto Jews began. At the border of the ghetto there was what was known as the *Umschlagplatz*, the trans-shipment place. Here were the railway sidings containing the box cars that carried waiting Jews to the death camps. Jews were forcibly driven to this assembly point. Any that resisted were killed on the spot. There were Jews who aided in this process in the hope that in this way they could buy a few more days or weeks of life. They, too, were eventually exterminated. Once the resistance in the ghetto got organized, some of these collaborators were executed by Jewish resistance fighters. Later the Germans brought in Ukranians to help with this roundup. Szpilman writes that these people were even worse than the Germans. Perhaps the most harrowing part of the film and the book is Szpilman's description of his family's displacement to the *Umschlagplatz*, which occurred on the August 16, 1942.

To picture the scene one must imagine people packed into a small area in midsummer with no food or water. There were bodies piled aboard with them, including children who had had their skulls smashed when they had been seized by the legs and swung violently against walls. The film leaves out this detail. It does include a woman who kept repeating "Why did I do it?" This refers to the fact that she had smothered her baby accidentally while trying to keep it from revealing her hiding place. As it was dying it made enough noise so that the Germans found the rest of the family. At one point, Szpilman's family was able to scrape together enough money to buy from a single cream caramel from a boy among the detainees who was selling at it a very high price. His father divided the candy into six parts.

(Under the circumstances, what the boy planned to do with the money is not clear.) It was their last meal together. Then it was their turn to be herded into the cattle cars. Szpilman writes, "We had gone half way down the train when I suddenly heard someone shout, 'Here! Here, Szpilman!' A hand grabbed me by the collar, and I was flung back and out of the police cordon."[92] Szpilman's reaction was to try to break through the wall of police and rejoin his family. He was able to shout to his father who had seen him, but he could not reach him. That was the last time he ever saw his family. One of the Jewish policemen said to an SS officer in German, "Well, off they go for meltdown."[93] To understand the rest of the book and the film one must have some understanding of the German invasion of Poland.[94] This you will not find in the film. This invasion was qualitatively different from, say, the German invasion of France or Denmark. These countries had cultures for which the Germans had some respect. Their goal, apart from pacifying the populations, was to persuade them of the superiority of Germany and the German way of life. With Poland something completely different was intended. As noted in the last essay, Polish culture was to be eliminated, and the Poles were to be reduced to drones, to be exported to Germany to work as slaves and then disposed of. Hitler put it clearly when he wrote, "It is essential that the great German people consider as its major task to destroy all Poles."[95] Hans Frank put it somewhat differently. He said, "You must not kill the cow you want to milk. However, the Reich wants to milk the cow...and kill it."[96] He managed to do both, and for this he was tried at Nuremberg and hanged. This German goal had the effect of radicalizing the Polish population and putting them in the same position as the Jews. There was, and is, a considerable undercurrent of anti-Semitism in Poland, but during the war much of this was muted in the face of a common threat. This may be one of the reasons why so many Poles risked their lives trying to save Jews and why some of them risked their lives to save Szpilman's.

After escaping from this deportation Szpilman returned to the ghetto, or what was left of it. He was too demoralized to try to flee, and to where? When he finally did escape, his escape was carefully orchestrated. Labor in the ghetto was organized by a Jewish Council who collaborated with the Germans until they, too, were exterminated. Through someone Szpilman knew on the council, he was assigned to various jobs involving manual labor, for which he was totally unsuitable. He managed to help smuggle arms into the ghetto and came very close to being caught. As the days went on, the Jews doing this work were divided into groups, some of which were slated for deportation. It became clear to Szpilman that it was only a matter of time before he, too, would be deported and that he must find a way to escape. In the film, Polanski gives the impression that one could simply walk out of the ghetto so long as one could

---

[92] *The Pianist*, p. 105.

[93] *The Pianist*, p. 107.

[94] An excellent source is Forgotten Holocaust by Richard C. Lukas, Hippocrene Books, New York, 1997.

[95] Lukas, p. 4.

[96] Lukas, p. 5.

meld into the general population. As I said earlier, any Jew caught leaving the ghetto was immediately executed. Escape required careful planning. There were couriers to the underground both inside and outside of the ghetto who were skilled at finding their way in and out. One of them contacted a couple—the Boguckis, he an actor and she a singer—to set up such a plan. Much is made of them in the film, although in the book they play a cameo-but-crucial role. What happened, Szpilman tells us, is that he was able to slip out of the ghetto by joining a group of non-Jewish workers who had been working in the same project just as they were leaving. Bogucki was waiting for him at a pre-assigned place and led him to an unused artist's studio where he could stay hidden, at least for a short while. He was ultimately relocated in a bachelor's flat, where he was brought food twice a week and also news that he got from the underground newspapers that flourished in Warsaw. It was from these that Szpilman learned about the uprising of the Jews in the ghetto that began on April 19, 1943, and lasted a month before it was put down. In the film, Polanski locates Szpilman's flat in sight of the ghetto so that he witnesses the uprising from close at hand. In the book Szpilman tells us that if he leaned out his window he could see the light from fires on the distant horizon in the direction of the ghetto. Polanski could not resist the temptation of showing Jews fighting back when they had the chance.

In August, Szpilman's hiding place was discovered by his neighbors, and he had to flee for his life. Once again he found temporary refuge with courageous Poles who risked their lives by taking him in until they found for him his last pre-arranged hiding place—a large flat in a quarter occupied mainly by Germans. Here he remained until August of 1944, when the Warsaw uprising began. The revolt had been in the offing for several years but held off to wait until the Russians were at the outskirts of Warsaw. The Jewish underground army consisted of about 25,000 men, of which only some 2,500 were armed. The Germans had about 16,000 well-armed troops, which were reinforced as the rebellion proceeded. By August 7 Warsaw was on fire. Hitler ordered the city to be reduced to rubble. Szpilman could watch some of this from his flat until on August 12 he learned that the Germans were about to shell the building. With his building on fire, Szpilman decided to kill himself by taking an overdose of sleeping pills, which he had been given by a doctor to alleviate the pain of a liver ailment he had contracted. He lay down on his sofa expecting to die. But he didn't.

The next morning he awoke and then struggled to get out of his burning building. In the total chaos he went to the now empty German hospital across the street hoping to find food and water. He managed to locate some filthy water in buckets that were to be used in case of fire and some moldy crusts of bread. It soon became clear that the building was not a safe place to stay since the Germans kept returning to search it. For several weeks he scavenged among the hulks of buildings until he finally found shelter in the ruins of a bungalow that had once been part of an estate. But here he was nearly apprehended and had to flee to a nearby villa, where he lived in attic. It was now November and, in addition to everything else, Szpilman was freezing. He was in the process of searching for food when, as he writes, "I was so absorbed in my search that I never heard anything until a voice right behind me said, 'What on earth are you doing here?'"

A tall elegant German officer was leaning against the kitchen dresser, his arms crossed over his chest. "What are you doing here?" he repeated. "Don't you know the staff of the Warsaw commando unit is moving into this building any time now?"[97] With this, the most extraordinary moment in the film and the book begins.Szpilman collapsed on a chair and said to the officer, "Do what you like with me. I'm not moving from here."[98] The officer assured him that he had no intention of doing anything to him and asked him what he did for a living. Szpilman said that he was a pianist. There then occurs something which is surely one of the great moments in film and in the book. The officer asks that Szpilman follow him to the next room, where there was piano, and demands that he play something. Szpilman is terrified that the SS will hear, but the officer assures him that he will tell them that he was trying out the piano. Szpilman sits down to the piano, which had been damaged by being exposed to the open air. He had not played for two and half years and his fingers were covered with a thick layer of dirt. He writes:

"I played Chopin's Nocturne in C sharp minor. The glassy tinkling sound of the untuned strings rang through the empty flat and the stairway, floated through the ruins of the city and returned as a muted melancholy echo. When I had finished, the silence seemed even gloomier and more eerie than before. A cat mewed in a street somewhere. I heard a shot down below outside the building-a harsh, loud German noise."[99]

The officer then offers to take Szpilman out of the city where he will be safer. Szpilman says he can't go. Now the officer realizes that Szpilman must be Jewish. He asks him where he had been hiding and accompanies Szpilman to his hiding place. He notices that above the attic there is a loft with a ladder that can be pulled up once one has climbed it. He tells Szpilman that this is where he must hide. He asks him if he had had anything to eat and when he learns that Szpilman has not eaten he offers to bring food. Szpilman is so surprised that he asks the officer if he is really German. He is told that he is and is ashamed of it. Three days later the officer is back with food. On December 12 the officer is back for the last time. He is leaving Warsaw with his detachment. Szpilman tells him that if he survives he will once again be working for Polish radio and he tells him his name. Polanski cannot resist a German pun and has the officer say "'Spielmann' is a good name for a pianist." This is not in the book and it is not even made clear whether the conversation took place in German. It turned out that this remarkable soldier had taught himself Polish. The last thing that Szpilman says to him is to remember his name in case there is anything he can ever do for him. They never meet again. In the book Szpilman says that he did not ask for the name of the officer because he was afraid that if he had been captured and tortured he might have given it away.

---

[97] Szpilman, pp. 175–176.

[98] Szpilman, p. 177.

[99] Szpilman, p. 178.

In the film we see a ragged column of Poles returning to Warsaw. They pass a field enclosed with barbed wire containing captured Germans. They spit on them, but a worn, ragged officer approaches the barbed wire and says that he had helped Szpilman and now needs his help. He tries to shout his name, but he cannot be heard. This is how the film ends except for the final concert.

There are a couple of captions. The first one notes that Szpilman died in Warsaw in the year 2000 at the age of 88. The second one reveals the name of the officer—Wilm Hosenfeld—and says that he died in a Soviet prison camp in 1952. It adds that nothing more is known about him. This statement is incorrect. A great deal is known about Hosenfeld. In fact in the book there is a section devoted to Hosenfeld's diary—a remarkable document—and there is an epilog by the German writer Wolf Biermann, who was responsible for the reissue of Szpilman's memoir in Germany. Biermann tells us a good deal about Hosenfeld, and there are numerous German language websites with a great deal more. He seems to have become something of a folk hero in Germany. Why Polanski chose to ignore all this is unclear. Even if he did not know some of it while he was making the film, this is something he could have amended later.

Wilm Hosenfeld was born on the May 2, 1895, in Rhoendorf Mackenzell, the fourth of six children in a very religious Catholic family. His father was a teacher. Hosenfeld himself served in World War I and was wounded three times. Like his father he, too, became a teacher, and in 1920 he married Annemarie Krummacher, who was the daughter of a painter. They had five children. In the film something that looks like a family photograph is shown on Hosenfeld's desk. It went by too fast for anyone to see how many children there were. He joined the Nazi Party in 1933 and became a Storm Trooper. When World War II came, he was considered too old for active service but was assigned to Warsaw as an *Oberfeldcommndantur* in charge of the sports facilities the Germans had seized. The extracts from his diary begin in 1942, when he is already thoroughly disillusioned by his fellow countrymen. (By the way, a photograph taken of him at this time shows a remarkably elegant looking man.) On January 18, 1942, he writes, "Hitler says he is offering the world peace, but at the same time he is arming in a disturbing manner. He tells the world he has no intention of incorporating other nations into the German state and denying them the right to their own sovereignty, but what about the Poles and the Serbs? Especially in Poland. There can be no need to rob a nation of sovereignty in its own self-contained area of settlement."[100]

By April he informs us that while he is spending peaceful days in the College of Physical Education, he is not happy. He is beginning to see the atrocities that are being carried out by his fellow Germans. It is noteworthy that he is aware of Auschwitz and the fact that people are being gassed there. An entry written on August 13, 1942, three days before Szpilman's family was deported, a few weeks after Hosenfeld had learned that Jews were being systematically exterminated in concentration camps, he writes:

---

[100] Szpilman, p. 194.

"You can't help wondering again and again how there can be such riff-raff among our own people. Have the criminals and lunatics been let out of the prisons and asylums and sent here to act as bloodhounds? No, it's people of some prominence in the State who have taught their otherwise countrymen to act like this. Evil and brutality lurk in the human heart. If they are allowed to develop freely they flourish, putting out dreadful offshoots, the kind of ideas necessary if the Jews and Poles are to be murdered like this."[101]

A year later he writes:

"These brutes think we shall win the war that way. But we have lost the war with this appalling mass murder of the Jews. We have brought shame upon ourselves that cannot be wiped out; it's a curse that cannot be lifted. We deserve no mercy; we are all guilty."[102]

But Hosenfeld was not content with Jeremiah-like lamentations. He began to act. His saving of Szpilman was not an isolated act. It was the last of several. Some are documented in Biermann's epilog. One of them involved a Polish Jew named Leon Wurm. In 1950, Wurm went to visit the Hosenfelds in West Germany. He learned from Hosenfeld's wife that, in 1946, she had received a postcard from her husband from a Soviet prison camp with a list of names of people who might help. Szpilman's was one of them. This was the first time that Szpilman learned the name of that German officer. He immediately went to the head of the Polish secret police, a man named Jakub Berman, whom he despised. He told Berman not only his own story but that of the others. Berman was impressed and tried to help, but the Russians would not let Hosenfeld go. On August 8, 1957, Hosenfeld, now suffering from extreme depression, died in the prison camp where he had been held. He was 57.

Szpilman resumed his career on Polish radio. He continued to compose music and resumed a musical partnership with the Polish violinist Bronislav Gimpel. In 1957, they toured West Germany and took the opportunity to visit Hosenfeld's widow. She gave him a picture of her husband. When Biermann was preparing the new edition of Szpilman's book he spoke to Szpilman about Hosenfeld. Szpilman said that he had never discussed it with anyone, including his wife and two sons. "Why not you ask? Because I was ashamed,"[103] ashamed that he had not succeeded in saving the life of the man who had saved his.

---

[101] Szpilman, p. 199.

[102] Szpilman, p. 205.

[103] Szpilman, p. 220

Chapter 13
Rocket Science

> *'They're falling in a Poisson distribution'* says Pointsman in a
> small voice, as if it was open to challenge.
> *No doubt man, no doubt—an excellent point. But all over the
> fucking East End, you see.*
>
> —Thomas Pynchon Gravity's Rainbow[104]

> *When he first met [Arnaud] Desplechin in 1982, [Michel]
> Djerzinski was finishing his doctoral thesis at the University of
> Orsay. As part of his studies he took part in Alain Aspect's
> groundbreaking experiments, which showed that the behavior
> of photons emitted in succession from a single calcium atom
> was inseparable from the others. Michel was the youngest
> researcher on the team.*
>
> —Michel Houellebecq, The Elementary Particles[105]

Not long ago I found myself in the Libreria Francese on the Piazza Ognissante in Florence. I was on something of a mission. I had watched a re-broadcast of the inaugural of the French cultural television program "Campus," which was the successor to the various very popular Bernard Pivot programs such as "Bouillon de Culture." The master of ceremonies was an attractive and intelligent young man named Guillaume Durand. For the inaugural Durand promised a treat, an interview with the "bad boy" and present sensation of the French literary scene, Michel Houellebecq. Durand explained what a rarity it was for Houellebecq to consent to an interview like this, and it soon became apparent why.

In person, Houellebecq resembled a somewhat oversized, pellucid rodent. He chain smoked throughout the interview, which made much of what he said unintelligible. The questions that Durand asked were worse than unintelligible. They were absurd-things, such as "Are you a Pétainiste—a racist?" "Do you think about what you write?", etc. In short, it was a fiasco. Nonetheless, something about Houellebecq

---

[104] *Gravity's Rainbow* by Thomas Pynchon, Bantam Books, New York, 1974, p. 200.

[105] *The Elementary Particles*, by Michele Houellebecq, translated from the French by Frank Wynne, Knopf, New York, 2000, p. 103. This is translated from *Les particules élémentaires*, Flammarion, Paris, 1998.

intrigued me. It was clear that behind the cloud of cigarette smoke there was an interesting mind at work. Thus, my mission. I had decided to read at least one book of Houellebecq's, and I was looking for advice as to which one to chose. The Frenchwoman who ran the bookstore suggested that I begin with *Les Particules Élémentaires*. She said it was the most accessible. So, following her advice, I bought the book.

I had no idea what to expect. I thought it might even be a popular disquisition on my field of physics, the theory of elementary particles. Of course, it turned out to be a novel, in turns brilliant, scabrous, and indeed downright pornographic—very funny and, in the end, tragic. One understands what the fuss was about. But after I had read some of it I had an odd sensation, a kind of *déjà vu*. Somewhere I had read this before, not exactly this, but something like this. Then it struck me. It was Pynchon's *Gravity's Rainbow*. It was not the plot. In so far as Pynchon's novel *has* a plot, its protagonist is the V-2 rocket, the ballistic missile that the German's aimed at London beginning on September 8, 1944, and ending in March of 1945, by which time some 600 had hit the city. Everyone in Pynchon's book is in one way or another related to the rocket. On the other hand, Houellebecq writes about two stepbrothers, Bruno and Michel, who do not meet until their teens. Bruno's father informs Janine, the boys' mother, that it is time for her two sons to meet, and they are introduced with hardly any preparation. Afterwards, they become friends. The book is about what happens to them. So there is nothing in the plots of the two books that suggests a resemblance.

It is true that both books are very colloquial. I do not think the slang in Houellebecq's book is really translatable, and having read both the French and English versions I strongly suggest trying to read it in French. In any event, using slang in novels is pretty common, so that was not it. Nor was the sex. Both books are drenched in sex. Indeed, after reading graphic descriptions of corophilia and urolagnia the only urge they gave me was to consider joining a Buddhist monastery. So it wasn't the sex. What was it then? It was the science. This requires some explanation.

It is not hard to find examples of science in novels. For example, C. P. Snow's novels often have scientists. But they never discuss science, and it plays no real role in the novels. If you replaced the scientists by professors of Sanskrit, nothing would be lost. There is, of course, science fiction. Like pornography, I find science fiction almost impossible to define, although I know it when I read it. It does involve science, but not real science. In the old days it was life on Mars with the Martians invading Earth. Now, some of the more outré interpretations of the quantum theory—parallel universes, for example—are the stage on which androids of various kinds perform. A friend of mine once complained that mixed doubles was neither sex nor tennis. Substitute "science" and "fiction" then, as far as I'm concerned, and you about have it. These two novels are not science fiction. It is my claim—and this what I hope to demonstrate—that they cannot be fully appreciated without the appropriate scientific culture.

When it comes to other kinds of culture we readily accept this fact and take, for example, whole courses on Proust's Paris or Joyce's Dublin. We can read their books without this background, but we miss a great deal. In the case of Pynchon and Houellebecq the novels can, and are, read by people who do not understand the scientific background, but they have no idea what they are missing. Let's illustrate this with two detailed examples, one from Pynchon and one from Houellebecq. The first is from Pynchon.

As mentioned before, the protagonist of Pynchon's novel is the V-2 rocket, "V" for *Vergelstungswaffwe*, a reprisal weapon. (In the novel it is sometimes referred to as the "A4," another designation used by the Germans.) This was a totally new instrument of war. The Germans began designing it in 1940 under the guidance of Wernher von Braun and General Walter Dornberger. By 1943, V-2s were being manufactured in a system of underground tunnels under the Kohnstein Mountain near Nordhausen. The tunnels were constructed by concentration camp inmates taken from Buchenwald. This subcamp of Buchenwald was called Dora—it is described in the novel—and the plant itself was called Mittelwerk. It is estimated that about 60,000 detainees were used to make the tunnels, of which some 26,000 did not survive. Each shift, men living under inhuman conditions, worked 12 hours. They were worked to death, and many of the survivors were slaughtered when, in April of 1945, the SS liquidated the camp. By that time, some 4,500 missiles had been produced.

The V-2 was an enormous rocket, about 46 feet long and 5 feet in diameter. It had a payload of something like 975 kilograms. Its fuel was a mixture of alcohol, water, and liquid oxygen. Using engines newly designed for the rocket, it had a range of about 200 miles and climbed to a height of 35 miles. It reached a maximum speed of about 3,300 miles an hour, several times the speed of sound, which dropped to about 1,800 miles an hour, still supersonic, when it impacted. This produced a very characteristic sound pattern. The first thing one heard was the sound that it made when it impacted the ground. Then, immediately after, there was the explosion. And finally, there was the sound —the whine and roar—that the rocket had previously made while it was moving towards the target. This was just the opposite of the V-1, or "buzz bomb," which was subsonic. You heard the V-1 coming. You knew you were in trouble when you stopped hearing them because it meant that the motor had shut off and that it was about to land on your head. Because it was comparatively slow, about 60 percent of the V-1s were shot down by airplanes or antiaircraft before they could explode. Nothing could shoot down a V-2. By the end, nearly 3,000 Londoners were killed and over 6,000 seriously injured. Pynchon proposes a little limerick:

> There once was a thing called a V-2,
> To pilot which you did not need to—
> You just pushed a button,
> And it would leave nuttin'
> But stiffs and big holes and debris, too.[106]

Now let us focus on the rocket's guidance system. This will lead us to our scientific example.

---

[106] Pynchon, op. cit., p. 355.

The V-2 was the first truly ballistic missile. This meant that its trajectory—the parabola it would follow (Pynchon loves parabolas)—is set before the rocket is launched. Once it is launched, it is on its own. To keep it from straying from its assigned trajectory the system that is employed is called "inertial guidance." One can imagine that a platform has been constructed on the rocket that always maintains its initial orientation no matter what gyrations the rocket makes. In practice, the platform was maintained by three gyroscopes which, as one knows, resist change in their orientation. While the motor was still running, the gyroscopes steered the rocket with a system of vanes and rudders.

Although the British did not know this, the accuracy of the guidance system was not very good. You could aim it for a target as big as London but not, say, for Buckingham Palace. At some point during the war, probably near the end of it, judging from the numbers used, it occurred to a British actuary named R. D. Clarke that the question of the accuracy of the guidance could actually be answered mathematically using data collected from the frequency and disposition of the rocket hits without knowing anything about the guidance mechanism. Clarke wrote a one-page paper entitled "An Application of the Poisson Distribution," which he published in 1946 in the British *Journal of the Institute of Actuaries*.[107] Not much is known about Mr. Clarke except that in his paper he indicated that he was of the Prudential Assurance Company, Ltd. He notes that he came upon this method of examining rocket strikes in the course of a recent "practical investigation." One wonders of what and when, but as mentioned, from the numbers he employed, it must have been near the end of the V-2 raids. In any event, this one-page paper is now a classic. It is a textbook example used in almost any course on statistics and probability. It is clear to me that Pynchon must have studied it, perhaps during the period at Cornell University when he was taking engineering courses.

To understand Mr. Clarke's paper we must go back to the nineteenth century and the work of the great French mathematician Siméon-Denis Poisson. Poisson worked in a number of fields, but at some point he decided to use statistics and probability to study the French judicial system. He wanted to see if there were biases in guilty verdicts in various courts or whether they followed the sort of random pattern that the laws of probability would predict. In 1837, he published a book with the incredibly cumbersome title *Recherches Sur La Probabilité Des Jugements en Matière Criminelle et en Matiére Civile, Precedes Des Règles Générales du Calcul des Probabilities*.[108] The part of the book that is still remembered has to do with a simplification he introduced in the calculation of these probabilities, and indeed all probabilities that fall into this domain. Prior to Poisson's work, the exact formula for computing these probabilities was well-known,

---

[107] R. D. Clarke, *Journal of the Institute of Actuaries*, Vol. 72, 1946, p. 481. I am grateful to the librarian of the Fine Hall library at Princeton for supplying me with this document.

[108] The book is available in a modern replica, Elibron Classics, 2003. Poissons' preface is very interesting. He explains his concern about the judicial system and where he got his numbers. The approximation now known as the "Poisson distribution" is introduced with no fanfare on p. 41.

but rather cumbersome. Poisson pointed out that if the number of events was in some sense large and the individual probabilities small, then this awkward formula had a very simple approximate form, now known as the Poisson distribution. It contains three elements. There is a constant, which we will call 'C,' that normalizes the expression. There is a parameter 'm,' called this by both Mr. Clarke and Pynchon, which characterizes the particular case at hand. How this works in practice will become clear when we discuss the rocket example. Finally there is an integer that can take the values 0,1,2,3…and so on. This integer, which we will call 'x,' defines the number of events that can take place coincidentally in whatever region of space or time we are considering. Again this will become clearer in the rocket example. The only other thing we have to know is the meaning of what is called the "factorial" symbol, which is denoted by an exclamation point '!'. You will see immediately what this means when we give examples. So $0! = 1$, $1! = 1$, $2! = 1 \times 2 = 2$, $3! = 1 \times 2 \times 3 = 6$, and so on. You will have no trouble in seeing that $4! = 24$, etc. If we want the probability that there is a coincidence of x events, in a situation characterized by the parameter m, then Poisson tells us that this is given by $C \bullet m^x/x!$. In other words, you raise m to the power x and divide by $x!$. This is the Poisson distribution. But what has it got to do with rockets?

What Clarke proposed was to take an area of 144 square kilometers in south London and divide it into squares of a quarter of a square kilometer each. This comes to 576 squares. He must have drawn these on a map, but he does not tell us that. These 576 squares become our universe. He wanted to see if the Poisson distribution would predict the number of squares in which there were no hits, one hit, two hits, and so on. These are the integers x in this case. It turned out that in the period studied there had been 537 V-2s that had landed in this domain. Thus the parameter m is given by 537/576, which is the overall ratio of the number of hits to the number of squares. The constant C is determined by the requirement that if you add up all the squares where there are no hits, one hit, two hits, and so on you recover the total number of squares-576.[109]

With this information Clarke was in a position to see what the Poisson distribution would predict and to compare this with what was actually observed. He gives a table in his paper, and it is amazing how good the fit is. For no hits, for example, the Poisson predicts 226.74 squares. We don't get an exact integer because these are probable numbers. We can compare this to what was observed; i.e., 229 squares. For two hits the Poisson predicts 98.54 squares while the number observed number was 93 squares. This kind of agreement cannot be accidental. What it showed was that the rockets fell more or less at random all over London and, despite appearances, were not clustered around special targets, which was a limitation of the guidance system.

---

[109] For the mathematically oriented this sum is given by $576e^{-m}(1 + m + m^2/2! + m^3/3! + …)$. The sum in brackets, which is a sum over no hits, one hit, two hits, and so on, adds up to the exponential e^m, which then cancels the exponential e^{-m}, leaving the 576 as advertised.

This seems to have been what impressed Pynchon, the element of chance. In the novel, Clarke becomes the statistician Roger Mexico. Mexico also makes a map of London and divides it into squares. A hit in a square was represented by sticking a round-headed pin into the map. For all we know, Clarke may have done the same. As the pins build up, Mexico sees the Poisson distribution unfold and realizes its significance. He feels like he is being accused of precognition, so he explains Poisson to anyone who will listen. One of his auditors is Edward W. Pointsman, Fellow of the Royal College of Surgeons. When we meet him he has just been bombed out by a V-2 and is being rescued by Mexico. Needless to say, he does not want another V-2 coming in his direction. Pointsman asks, "'Can't you...*tell*...from your map here which places would be the safest to go into, safest from attack?' Mexico's answer is "No."

"But surely—

"Every square is just as likely to get hit again. They aren't clustering. Mean density is constant.

"Nothing on the map to the contrary. Only a classical Poisson distribution, quietly neatly shifting among the squares exactly as it should...growing to its predicted shape...

"But squares that have already *had* several hits, I mean—

"I'm sorry. That's the Monte Carlo Fallacy. [If the wheel is fair the chance of black coming up in the next spin is the same no matter how often black has come up before.] No matter how many have fallen inside a particular square, the odds remain the same as they always were. Each hit is independent of all the others. Bombs are not dogs. No link. No memory. No conditioning."[110]

There was something democratic about these bombs. They fell on the rich and the poor, the mighty and the humble, with equal probability. I kept thinking as I read Pynchon that there was something naïve, almost romantic, about all of this. With a modern hydrogen-bomb-tipped ballistic missile, one would be enough to take out the whole city. No Poisson distribution here.

Now to Houellebecq and a totally different science.

The younger of the two brothers in Houellebecq's novel, Michel Djerzinski, is a scientific genius. This first, as it often does, manifests itself in mathematical precocity. Then Michel studies physics at the Orsay campus of the University of Paris. In 1960, when I was doing physics in Paris, our group was transferred to Orsay, a Parisian suburb. The campus was still unfinished, and, as I recall, there was a good deal of mud. People were just getting organized. But this part of Houellebecq's novel is set in the early 1980s, by which time Orsay was thriving.

While there, Houellebecq's Michel takes part in an experiment done by the French physicist—a real French physicist—Alain Aspect. Since this experiment was part of Aspect's 1983 Ph.D. thesis it is unlikely that our paths crossed in 1960. It is such an interesting choice for Houellebecq to have made for his fictional character that this is what we shall focus on. What is the experiment, and why the

---

[110] Pynchon, op. cit., p. 64.

choice? But first to complete my account of the novel. Michel's abilities are recognized by a molecular biologist named Desplechin, who invites Michel to work in his laboratory. In the course of time, Michel has a biological insight, which would lead to a species that can reproduce asexually. I am not enough of a biologist to deconstruct this aspect of the novel. My guess is that it is largely fantasy. On the other hand, I can adumbrate the physics. Let's begin with Bertlmann's socks.

Reinhold Bertlmann is an Austrian theoretical physicist who often visited the elementary particle laboratory CERN in Geneva. One of his eccentricities caught the attention of my friend and colleague the late John Bell. Bell was an Irish-born physicist at CERN, who died in 1990 of a sudden stroke at the age of 61. Bell noticed that Bertlmann never wore socks that matched. Inspired by this observation, Bell wrote a wonderful article that he called "Bertlmann's Socks and the Nature of Reality."[111] In his article, which was widely read, Bell managed to describe all the issues involving the interpretation of quantum mechanics—the whole Bohr-Einstein debate—issues that Bell had been largely responsible for reviving interest in and for which work he had been nominated for a Nobel Prize. As we will explain, Aspect's experiments are a very important part of this story, but back to the socks.

As Bell noted, if you happened to observe Bertlmann wearing a pink sock you could be absolutely certain that the other sock was not pink. On the occasion of the memorial symposium held for Bell at CERN, Bertlmann had on, as I recall, red and green socks.

Now let us extend the sock image as follows: Let us suppose that Bertlmann's socks were rigorously color coordinated. That is to say, if Bertlmann wore a green sock, the other would be guaranteed to be red, pink would go with white, blue with orange, and so on. Let us imagine that somehow we manage to persuade Bertlmann, for the sake of an experiment, in our presence to exchange say a red sock for a blue one. Common sense tells us that the other sock is not going to turn orange. That would violate our notions of the objective reality of socks and, at the very least, we would demand an explanation. What I am now going to try to convince you of is that on the subatomic level this kind of *leger de main* actually takes place and that, if one believes quantum mechanics, there is no explanation, if by explanation you mean a causal sequence of events of the kind we usually mean.

As a warm up exercise, so to speak, let's start with something that is already pretty bizarre but not quite Bertlmann. Particles such as electrons, protons, or atoms have, in addition to a set of properties we would expect, such as mass and electric charge, another property we would not expect unless we had taken a course in modern physics—spin.

Spin is a kind of angular momentum. It differs, however, from the angular momentum that Earth has going around the Sun, or that an object on a string would

---

[111] The essay can be found in Bell's collection *Speakable and Unspeakable in Quantum Mechanics*, Cambridge University Press, Cambridge, 1987.

have if we whirled it around our heads. This angular momentum is properly called "orbital angular momentum." It vanishes if the object is brought to rest. But even a resting object can have an angular momentum if it is spinning around like a top. If we don't push the analogy too far, spin can be thought of as something like this sort of angular momentum. When it comes to particles, physicists refer to this as the "intrinsic" angular momentum of the particle. But these spinning particles act like tiny magnets with their north and south poles oriented in relation to the spin of the particle. If we put one of these tiny magnets in a uniform magnetic field both its north and south poles have the same force acting on them, but pulling in opposite directions. Therefore the trajectory of the particle is not deviated. On the other hand, if we replace the uniform field by one that is non-uniform, the north and south poles will feel different forces and there will be a net tug that will influence the path the particle follows.

In 1921, the German physicist Otto Stern proposed an experiment to check this out. As it happened, Stern had no experimental skills whatsoever (he never touched any of them), so he persuaded a very competent experimental physicist named Walther Gerlach to do the actual measurements. Stern suggested using neutral silver atoms for his test particles. By neutral here, we mean electrically neutral. This was important because with an electrically charged particle, any stray electric field will perturb its trajectory, obscuring the magnetic effect. What Stern thought would happen is that the tiny magnets associated with the silver atoms would be randomly oriented with respect to the non-uniform magnetic field he was employing. This meant, he thought, that if you collected these atoms on a glass plate they would form a kind of line since each atom would arrive at a different place because of the random orientation of its tiny magnet with respect to the external magnetic field. When the experiment was first performed, they saw nothing. But Stern was a cigar smoker, and the cheap cigars he could afford had a high sulfur content. The sulfur from Stern's cigars blackened the silver, and then they could see something, but what they saw seemed to defy explanation. Instead of the straight line they expected, all the silver was deposited in two separate lines. This is beyond the explanation of classical physics. The interpretation we give this is that for the silver atom the spin has only two possible orientations with respect to the fields—either parallel to it (or anti-parallel to it) "up" or "down with respect to the fields." If you like, the orientations are "quantized." People used to call this "space quantization," but it is not really space that is quantized. It is the spin angular momentum. This was the first experimental demonstration of quantum mechanics in the raw. Now, with this preliminary, we can return to Bell, Bertlmann, and Houellebecq.

Bell was born into a Protestant family in Belfast of modest means. None of his family had had a higher education. It was assumed that after the age of 14 the children would go to work. But Bell's brilliance in school was soon evident, and enough money was scraped together from somewhere so that he could go to a technical high school to learn a vocation. Bell found that he actually enjoyed working with tools and the like. When he graduated, there was again no money to go on, so Bell found a job at Queen's College setting up demonstration equipment for the laboratory courses. He made a great impression on the professors who employed him, and with their encouragement and the money he had saved, after a year he was able to enroll as an undergraduate at Queens. He took first-class honors in both experimental and theoretical physics. He also began his study of quantum mechanics. I once asked him if when he began studying the theory he thought that it might actually be wrong. He replied in his lilting Irish accent, "I hesitated to think it might be wrong, but I *knew* that it was rotten. That is to say, one has to find some decent way of expressing whatever truth there is in it."

I can still hear the relish in Bell's voice when he said "rotten." All the arguments he was offered seemed to him to be defective and illogical, an attitude that did not sit well with some of his professors. When he graduated from Queens he got a job with the British Atomic Energy Research Establishment working on the design of accelerators. There he met his wife Mary, who was also a theoretical physicist working on accelerator design.

In 1960, the Bells decided that there was no university in Britain that could accommodate both of them professionally, so they took jobs at CERN and moved to Geneva. That is where and when I first met them. They were engaged in designing the new CERN particle accelerator and Bell, who in the CERN roster always identified himself as a "quantum engineer," was doing conventional theoretical physics as well. What he was not doing in a serious way was an examination of the foundations of quantum mechanics.

This had largely to do with his work ethic. He felt he was being paid to design accelerators and to do conventional theoretical physics and not to explore questions that most people thought had more to do with philosophy than physics. But in 1963, the Bells were invited to spend a year at Stanford University to work on anything they wanted. Bell now turned his attention to the foundations of quantum mechanics. He began by studying a thought experiment that had been suggested by the physicist David Bohm. In 1951, he published an influential text on the quantum theory. I used it to learn the theory. At the time it was published, Bohm had gone to Brazil to escape the political atmosphere in the United States, which was becoming more and more McCarthy-like. He later moved to Britain, where he spent the rest of his career. The experiment that Bohm suggested in his text—remember that it was a thought experiment—was to imagine two Stern-Gerlach magnets located far away from each other. In our imagination we can suppose that the space between them has been evacuated so that there are no effects of stray electrical fields and the like. We suppose that each of these magnets has been aligned so that the magnetic fields they produce are in the same direction. They are parallel. In between them is a source of particles, say electrons. The source produces these particles in pairs. The individual electrons move off in opposite directions towards the waiting magnets. But we fix it so that the source prepares the electrons in a special spin state. The state is such that if one magnet detects a spin, say up, the other magnet is sure to detect a spin down, and vice versa. It is like Bertlmann's socks. The correlation is perfect. Quantum mechanics allows us to do this with electrons, but there is a profound difference between them and the socks.

With the socks we can demand an explanation of how the correlation was achieved. We can watch Bertlmann choose his socks and put them on one at a time. In quantum mechanics we cannot do the equivalent with the electrons. If we ask by what mechanism in quantum theory all the electrons at a given magnet know enough to line up parallel or anti-parallel we cannot supply one. On this matter the theory is mute. Different attitudes towards this are possible. Einstein felt that it showed that the theory was incomplete. It allowed unexplained what he called "spooky actions at a distance." It is often said that Einstein's complaint about quantum theory was that it dealt in probabilities—in God's playing dice. That was an early attitude, but as his ideas evolved what Einstein came to object to was the inability of the theory to answer what he felt were reasonable questions. It did not adequately describe "reality," he thought. Niels Bohr, who had done as much as anyone to develop the standard interpretation of the theory, felt that this situation was quite reasonable. He once wrote, "There is no quantum world. There is only an abstract quantum mechanical description. It is wrong to think that the task of physics is to find out how Nature is. Physics concerns what we can say about Nature."[112]

Although Bell acknowledged that Bohr was a great physicist, he felt that the things that he wrote about the interpretation of quantum mechanics were totally opaque. He kept referring to Bohr as an "obscurantist." On the other hand, he felt Einstein was totally lucid. In the debate that took place over three decades between Einstein and Bohr, Bell was rooting for Einstein. In fact, he decided to try to give Einstein a hand. Bell thought that he would look for a mechanism—a hitherto unsuspected force—that would cause the little spin-magnets to line up the way quantum mechanics said they would. He did not expect that this force would be conventional, but that was beside the point. It would be a mechanism. In fact, Bell found that it was very easy to invent such a force in the case at hand, when the two Stern-Gerlach magnets were lined up so their magnetic fields were parallel.

It was a pretty ugly-looking force, but it worked. Score one for Einstein. But then Bell tried something that no one else had thought to do. He said, suppose while the twin particles are moving towards their respective magnets, I rotate one of the magnets so the fields are no longer parallel. I change the color of one of Bertlmann's socks. What then? Quantum mechanics gives a clear answer. The correlations change. Bertlmann's other sock has changed color. Now we no longer inevitably have the spin up, spin down correlation. In a certain fraction of the cases, depending on how much we rotate the magnet, we can have, for example, spin up, correlated with spin up. This is quantum mechanics with its spooky action at a distance. But is there a force that would duplicate this result and satisfy Einstein's request?

Bell tried but he could not find one. He then began to suspect that something very deep was going on here. He began to suspect that there was no such force and that this offered a clear choice between Einstein and Bohr. Remember we are talking about imaginary experiments so far, so we can't appeal to experiment to settle this. What Bell showed is that assuming such a force exists, the correlations would

---

[112] This quote can be found in M. Jammer, *The Philosophy of Quantum Mechanics*, Wiley, New York, p. 204.

necessarily obey a mathematical relationship now known as a Bell inequality. If this inequality was not obeyed, Bohr was right and Einstein wrong. It offered the first opportunity to decide between these two titans on the basis of an experiment—if one could actually carry out such an experiment.

Bell chose to publish his result in a journal called *Physics*, which did not last much longer than the issue that contained Bell's article. He made this choice because the journal did not assess page charges to contributors, which many scientific journals do. He felt that since he was a guest at Stanford he had no right to stick the Physics Department there with the cost of publishing his article. In fact, *Physics* actually paid the authors of the articles they published, which may be one of the reasons they went out of business.

The article appeared in 1964, and for the next five years no one paid much attention to it. But in 1969, four physicists—John Clauser of Berkeley, Michael Horne and Abner Shimony of Boston University, and Richard Holt of the University of Western Ontario—published a paper that changed everything. These physicists noticed that you did not need electrons to do Bell's experiment. You could do it with light. Light quanta—photons—also have a spin. This is not associated with a tiny magnet but rather with the polarization of the light. When light propagates, the light waves oscillate in a direction that is at right angles to the direction of propagation. This oscillation direction is called the direction of polarization. One can demonstrate it empirically by passing light through a polarizer. Now, if we emit two photons that are prepared suitably, their polarization directions will be correlated in a similar fashion to the spins of the Stern-Gerlach electrons. And there will be a Bell inequality. This was something that seemed possible to measure, which is exactly what Alain Aspect and his collaborators did in the early 1980s. They used photons from the radioactive decay of calcium and in a tour de force series of experiments showed that the Bell inequalities were violated. Bohr was right and Einstein was wrong. Here is how Houellebecq describes this in his novel.

"Aspect's experiments—precise, rigorous and perfectly documented—were to have profound repercussions in the scientific community. The results, it was acknowledged, were the first clear-cut refutation of Einstein, Podolsky and Rosen's objections when they claimed in 1935 that "quantum theory is incomplete." [In this paper, which was entitled, "Can Quantum Mechanical Description of Physical Reality be Considered Complete," Einstein, Nathan Rosen and Boris Podolsky proposed a much more complicated version of correlation experiments. Unlike Bohm's these did not involve spin.] Here was a clear violation of Bell's inequalities—derived from Einstein's hypotheses—since the results tallied perfectly with quantum predictions. This meant that only two hypotheses were possible. Either the hidden properties that governed the behavior of subatomic particles were non-local, meaning that they could instantaneously influence on another at an arbitrary distance [in making his argument Bell assumed that the forces he was considering did not have this non-local character, and most physicists would regard non-local forces as a kind of abomination], or else the very notion of particles having intrinsic properties in the absence of observation had to be abandoned. The latter opened a deep ontological void, unless one adopted a radical position and contented oneself with developing a mathematical formalism that predicted the observable and gave up on the idea of an underlying reality. Naturally it was that option that won over the majority of researchers."[113]

Houellebecq, I believe, has summed up the situation admirably. How he came to this understanding we do not know. One thing we do know is he did not discuss it with Aspect. Aspect informed me of that and added that the publisher never even sent him a copy of the novel. If Guillaume Durand had had a better understanding of what was involved he might have asked about the scientific background, rather than trying to find out if Houellebecq was a Pétainist—whatever that could possibly mean in the twenty-first century. He might also have asked why Houellebecq chose this set of experiments to have his protagonist Dzerzinski be a part of. He could have chosen any experiments, anywhere, After all, he was writing a *novel*. Why this one?

Here is a suggestion. It is inspired by something that Bell used to do in his popular lectures. Bell had spent time perusing the annals of an entity called The Institute for the Study of Twins. The institute had found two identical twins that had been separated as babies and had spent no time in each other's company. But they had an amazing list of coincidences. They smoked the same brand of cigarette, drove the same model car and bought it in the same color, and they both went to the same resort in Florida with their families. They both had dogs with the same names. Bell used to show a picture of them in his lecture—an example of spooky actions at a distance if there ever was one—unless, of course, you believe that a common set of genes explains it all. Houellebecq's novel deals with two brothers who never knew each other until many years after their birth. They influence each other even when apart, like the correlated electrons.

Of course, there is no straight line you can draw between an appreciation of quantum mechanics and a novel like Houellebecq's, any more than you can draw a straight line between the sight of the Moon and Beethoven's *Moonlight Sonata*. You are dealing with the workings of the creative imagination. What is interesting to see is how in both the creative imaginations of Pynchon and Houellebecq science played an important role.

---

[113] Houellebecq, op. cit., p. 103.

Chapter 14
The Science of Michel Thomas

When Michel Thomas published his first poems in *La Nouvelle Revue de Paris* in 1988, he adopted the last name of his paternal grandmother, Houellebecq.[114] He also changed his date of birth. Michel Thomas was born on the February 26, 1956, in Saint Pierre de La Réunion, a mountainous volcanic island in the Indian Ocean. His mother, Janine Ceccaldi Thomas, a medical doctor practicing on the island, had once made a visit to the cathedral at St. Michel and decided that if she had a son that is what he would be called. Michel Houellebecq, on the other hand, always gives his birth date as February 26, 1958, probably because, at least in the beginning, he wanted to keep entirely separate his literary career from what he was actually doing to earn a living. Michel Thomas was a computer systems engineer; he worked first in private industry, and then, from 1985 to 1991, in the Ministry of Agriculture until 1996, when he became an employee of the National Assembly. He was considered to be a very competent employee and when he left the National Assembly to write full time, he was told that his job would remain available to him until 2008. Since he is now a multi-millionaire thanks to his novels, it does not seem likely that he will take advantage of this offer.

In his novels Houellebecq's references are always very precise. It is not surprising, given his background, that this would apply to things involving computers. For example, in his latest novel, *La Possibilité D'une Île*,[115] the leader of a sect known as the élohimites, the "prophet," produces a website using HTML.[116] This stands for "hypertext markup language" and is the code one would use for preparing documents for the worldwide web. Needless to say, Houellebecq does not explain this. You either know it or you don't.

Michel Thomas's father René quit school at the age of 13 and after a brief career as a pastry chef became a mountain guide. He established himself in Val dèIsère. A guide friend of mine who knew him described him as "*assez marginal et condsidéré*

---

[114] An invaluable source is *Houellebecq, Non Authorizé* by Denis Demonpion, Maren Sell Éditeurs, Paris, 2005. I will refer to this as NA.

[115] Fayard, Paris, 2005, to be referred to as PI.

[116] PI, p. 250.

J. Bernstein, *Physicists on Wall Street and Other Essays on Science and Society*,
© Springer Science + Business Media, LLC 2008

com tel. Sympatique, mais assez à l'écarte.[117] "Sympathetic but rather aloof" is certainly how his contemporaries described Michel Thomas. Many people who have encountered Michel Houellebecq would leave out "sympatique." The senior Thomas met Janine in 1951, when she was part of a student group that used to practice rock climbing in the forest of Fontainebleau. They were married in 1953. They began traveling in various mountain regions and, after Michel was born, left him in Alger, in the care of his Ceccaldi grandparents, where he lived until 1961. By this time, the marriage was over, since Janine had had a daughter by another man. Although Michel visited his mother in La Réunion he never met his sister, who had been given to a local fisherman to raise, until she was 14 and he 18. They met in Holland for dinner. It was the one and only time they met. All of this has its echoes in Houellebecq's *Les Particules élémentaires*, in which his mother enters as a character in her own name. In the novel, as we have seen, there are two step-brothers who do not meet until their teens, but they do become involved with each other's lives.

In 1961, Michel's father more or less forcibly removed him from the Ceccaldi household. He had decided that Algeria was too dangerous. He installed Michel with Henriette, né Houellebecq, who was living with a man in Dicy in the Yonne. Houellebecq has said that the only decent member of his family was his grandmother. She was devoted to him and gave him something like a normal family life. He was a very precocious child and always first in his class. He was a natural to enter the Grandes écoles, the elite French secondary school system, which pretty much guarantees its graduates careers for life. To get into such a school the candidate must do two years in the Classe Préparatoire preparing for the qualifying examination. During those two years the students are stuffed with physics and, above all, mathematics, like Strasbourg geese. Some of my physicist friends went through this, at the end of which they were totally exhausted. You have to be very smart and very fast to survive.

In 1974, Thomas entered the lycée Chaptal in Paris to begin his two-year preparation. For no reason he could later give, he chose to orient himself towards biology, which did not mean that he did not go through the same mathematics program. After completing the course he took the examinations for two of the schools—the école Normale Supérieure and the Institut nationale agronomique, or the INA—which trained agricultural engineers. Normale "Sup" (the ENS) was, and no doubt still is, the most prestigious of the Grandes écoles. Thomas might well have gotten into it, having done well in one of the qualifying examinations (the literary one), but he decided he wanted to go to the INA, which was much less prestigious. Why, I am not sure. If he had gone to the ENS, which was a ticket for a political or executive career, one can only wonder if he ever would have become Michel Houellebecq.

Students at the INA, which he entered in 1975, were required to choose a specialty, and Thomas chose ecology. They also were required to put in time on a farm to gain practical experience. Thomas did well in the theoretical classes but was a

---

[117] Thanks to Claude Jaccoux for this observation.

disaster on the farm, and the farmer was glad to get rid of him. He also had no interest in sports, which were much in vogue at the INA. Nonetheless, his classmates found him agreeable, even interesting, but a little odd.

One thing he did at the INA was to produce a science fiction film using his classmates as actors. The year he graduated, 1978, was also the year his grandmother died, which for him was deeply traumatic. One wonders if he ever got over it. Naturally, he expected with his degree and his reasonable knowledge of German and English that he would get job offers as an agricultural engineer. For the next three years, despite constantly submitting his resume, he got none. He was officially "unemployed." He was able to get by very cheaply because his father had a studio in Paris in which he allowed his son to live.

Finding no employment as an agricultural engineer, Thomas decided to go to film school while still looking for employment as an agricultural engineer. This also required a competitive examination. He was selected, and spent two years from 1979 to 1981 in the école Louis-Lumière in Paris. He was able to see all the films he had heard of but had never seen. His professors and fellow students did not know quite what to make of him. He had a very strange background—agricultural engineering—and it was not clear to them why he was studying film-making. After he became Michel Houellebecq his teachers had difficulty making the connection. In 1980, he got married. His wife was from an upper middle class family who did not think much of Thomas. He did not seem to have any real profession. The next year the couple had a son, but the marriage was on the rocks. His young wife simply could not cope with a ménage involving a husband and a child. They divorced and Thomas began his second career as a computer technician.

Houellebecq has said that, since the age of 13, he knew he could write. He published his first prose work in 1991. It was a monograph on the American horror and science fantasy writer H. P. Lovecraft. This work was recently re-issued in translation under the title "H. P. Lovecraft: Against the World, Against Life," with an introduction by Stephen King.[118] Houellebecq came across Lovecraft when he was 16 and was apparently much taken by Lovecraft's nihilistic view of the world. Lovecraft inserted bits of modern science into his stories, which also must have appealed to Houellebecq. He was willing to accept as part of Lovecraft's literary persona his racism, his anti-semitism ("rat faced Jews"), and his admiration for Hitler, all of which are blatant in his writing. Lovecraft was himself profoundly neurotic and self-contradictory. For example, he had a brief marriage with a Jewish woman. That Houellebecq chose such a curious figure to write a monograph about raised questions about his own attitudes, which have never been satisfactorily answered. At the time that this monograph was published, Houellebecq had started writing his first novel, really a novella, entitled "*Extension du Domaine de la Lutte.*"[119] For some two years he tried unsuccessfully to get it published. It was turned down by all the known commercial publishers who must now have regrets. It was also turned down by

---

[118] Believer Books/McSweeny's, 1905.
[119] Maurice Nadeau, Paris, 1994.

Maurice Nadeau, a legendary publisher who introduced writers such as Georges Perec and Roland Barthes, as well as publishing people such as Samuel Beckett and Henry Miller. Nadeau was born in Paris in 1911, but still seems to be going strong. After a career in academics he worked as an editor and literary critic. It was in 1979 that he founded his own publishing house. In an interview he gave in 2002 in *L'Express Livre*, Nadeau noted that Houellebecq, "*m'a tanné pendant un an pour que je publie son premier roman.*" In this context I would translate *tanné* as "bugged," bugged for a year. One of the arguments Houellebecq gave was that he, Houellebecq, was the Georges Perec of his generation. While it is too early to place Houellebecq in the canon of contemporary fiction, the last writer that comes to mind when one thinks of him is Georges Perec. Anyone who has read, for example, *La Vie Mode d'Emploi* cannot help noticing Perec's sympathetic view of humanity in all its odd forms. A sympathetic view of humanity is about the last thing that comes to mind when one reads Houellebecq. Houellebecq apparently heard Perec lecture when Houellebecq was a teenager. But Perec died in 1982, so they never met.

In the same interview, Nadeau said of the Houellebecq novella he published that it was "*à peut près potable.*" "*Potable*" is usually an adjective applied to drinking water. One would love to know in what form Nadeau first saw the novella and whether the fact that it was nearly *potable* was a function of his editing. The novella is in the first person. The narrator is a computer engineer who works for a private consulting firm. No surprise there. One of its clients is the Ministry of Agriculture. No surprise there either. The narrator's—i.e., Houellebecq's—observation of both his colleagues and clients are shrewd, funny, and damaging. Apparently they were taken from life. The subjects had no idea that they had a sponge-like serpent in their midst. They also do not seem to have had much of a literary bent because few, if any, of them seem to have read the novella until it was pointed out to them that they were in it. Some were not happy. There are in the novella hints of graphic sex, which are a feature of Houellebecq's other novels. There are also hints of the dark side. The narrator throughout most of the novella seems to be the most lucid character in it. But towards the end, he has a nervous breakdown and commits himself to a mental institution. The novella ends when he is released to an uncertain future.

There is very little of the ontological baggage one finds in the later novels and there is essentially none of the science, which is what we will talk about here. The novella enjoyed reasonable commercial success. What happened next is subject to interpretation. It seems as if Houellebecq had a contract with Nadeau that obliged him to give Nadeau the option on his next book. The next book he proposed was a collection of poems that Nadeau turned down. He later claimed that Houellebecq had deliberately given him a collection of bad poems so he could get out of the contract. In any event, Houellebecq signed with the large commercial publisher, Flammarion. They got the gold ring. One cannot feel too sorry for Nadeau. The novella's commercial value has risen with the tide of Houellebecq's subsequent work.

Although I myself have studied and worked in France, I now live in New York, and so I have been out of touch with the latest French cultural currents. Indeed,

I had never heard of Houellebecq until, as I explained in the previous essay, I accidentally encountered him on the re-broadcast on a local New York television station of Guillaume Durand's interview. Houellebecq was then 45 and, as I previously remarked, resembled a pellucid rodent. Of his looks, Houellebecq once remarked, *"Dans la vie il y a les beau, il y a les moches. Et puis, il y a les gens comme moi. Ni beaux, ni moches."*[120] "In life there are the handsome ones and the ugly ones, and then there are people like me—neither handsome nor ugly." That fairly summed it up. As I have mentioned, the interview was pretty unenlightening. It was very early in Houellebecq's career, so I imagine he thought that he could use the publicity. What was remarkable to me was that this rather ordinary looking and not very articulate young man had written a novel, *Les Particules élementaires* that had become not only a best-seller, but a literary event.

As I discussed in the last essay, the novel is about two stepbrothers, Michel Djerzinski, born in 1958, and Bruno Ceccaldi, born in 1956. They had a mother in common, Jeanine Ceccaldi, who is a doctor. The fictional Michel was born while the fictional Jeanine was still married to Serge Ceccaldi, a plastic surgeon. Later she conceived Bruno while her husband was on a trip. The Ceccaldis were then divorced. Michel was sent to Yonne to live with his paternal grandmother while Bruno was sent to Algeria to live with his maternal grandparents. They do not meet until they are teens. Here, at once, we as readers of this story are confronted with a question. How important is it to know that Houellebecq-Thomas has divided himself into two and is, in part, describing his own life, having given his fictional mother his real mother's name and profession, even giving the two bothers his own two birthdates? Certainly we can read the novel without knowing this, but knowing it gives us an additional insight into Houellebecq's creative process. It is certainly not helpful when a critic such as Mark Lilla, writing in the *New York Review of Books*,[121] only notes in reference to the name Jeanine Ceccaldi that "many of his characters have foreign-sounding names." Foreign to whom? He thinks that Houellebecq might be trying to convey the notion that France is becoming a "mongrel nation."[122] It is to Houellebecq's science that we now turn.

By this time, the reader may well begin to wonder if it is necessary to know the underlying science to read Houellebecq's books. My answer is yes and no. They have been best sellers with sales of hundreds of thousands of copies in several languages. I would conjecture that 99.999 percent of its readers had no idea whether Alain Aspect was a real scientist or a figment of Houellebecq's imagination. Since the book is a work of fiction, it could be either. One can read the book without knowing. One can also read Proust without knowing that his fictional Balbec is actually the Normandy seacoast town of Cabourg, and that his Grand Hotel is still the Grand Hotel with its Proustian memorabilia. But, I submit that knowing this makes reading Proust's novel a richer experience. Likewise, knowing the science in

---

[120] NA, p. 195.

[121] Volume 47, Number 19, November 30, 2000.

[122] New York Review of Books, op. cit.

Houellebecq makes reading his novel a richer experience. The same is true of Pynchon.

In 2001, Houellebecq published his next novel, *Plateform*, also with Flammarion. In my view this was a novel in search of a good editor. About 25 percent of it is given over to repetitive, gratuitous descriptions of sexual acts. It is one of the few novels I have ever read where one looks for the clean bits. Also the novel, which takes place in Thailand where Houellebecq lived while he was writing it, is, as I will explain, outdated. It is narrated mostly in the first person by a Michel—another Michel—Renault, whom we meet just after his father has died. Michel, who works at the Ministry of Culture, decides to go on an adventure travel trip to Thailand. It is typical of Houellebecq that he would have Michel book his trip with the best known adventure travel company in France, Nouvelles Frontières. In the days when I was trekking in Nepal, we constantly ran into Nouvelles Frontières' groups. You could tell them because of the logos they wore and also because they all had almost the same equipment. This was not too surprising. They all had the same list, and they all bought what was on it at one of the Au Vieux Campeur sporting goods stores. So did I, by the way, and so do the people on Renault's tour. On the tour he meets a woman named Valérie, who later turns out to be a Nouvelles Frontières' executive.

What makes the novel dated is that all the beach resorts the tour goes to were washed away by the tsunami. It is hard to take seriously the descriptions in the novel of both the people and the scenery. In any event, after returning to Paris the two of them have a love affair, which gives Houellebecq license to describe their incessant copulations down to the last millimeter. Valérie and her boss leave their jobs to go to work for an even larger travel agency much more concerned about its financial returns. Michel has the brilliant idea that they should offer sex tours to places like Thailand. This was going like a house-a-fire when Muslim terrorists attacked the hotel where one of the groups was staying along with Michel and Valérie. Valérie is killed. This gives Houellebecq the opportunity to say a few negative things in general about Muslims. In that context, a context he invented, they are perhaps understandable. And if he had left it at that, probably nothing much would have been made of it. But he chose to go further with this. In the summer of 2001, he gave an interview in the literary magazine *Lire* in which he said, "*La religion la plus con, c'est quand même l'slam,*" "The stupidest religion is in any case Islam."[123] This created a furor, and an association of Muslim groups sued him for defamation. Houellebecq was then a legal resident of Ireland, but if he were to lose such a lawsuit he might be fined heavily, and this would come out of his royalties. He returned to Paris to defend himself in the judicial procedure. It became something of a cause among some French writers, who defended his right to self-expression. In the end he was let off with some cautionary remarks by the judge. His house in Ireland was vandalized, and he moved to an apartment in Dublin. While this was going on, his

---

[123] NA, p. 328.

publisher Flammarion decided that it needed to act. It sent a representative to the Grand Mosque in Paris to apologize. This infuriated Houellebecq and was the beginning of the end of his relationship with Flammarion. His latest novel is published by Fayard.

La Possibilité D'une Île is Houellbecq's most complex novel. In a vague sense it continues a theme of *Les Particules élementaires*. Bruno, the older of the two stepbrothers, at the end of that novel, commits himself to a mental institution where he remains. Michel, the other brother, has left physics to become a molecular biologist. He also decides to leave France and join a research institute in Ireland. He has, by some opaque mathematical reasoning, come to the conclusion that sexual reproduction is what renders us mortal. He seems to have figured out a genetic engineering way to avoid this and to achieve immortality. The science here is rather fuzzy, at least to me. And we never learn really what it is. Michel writes notes on it, which he leaves after he apparently commits suicide. Only then are they discovered and appreciated and his genius recognized. In *La Possibilité D'une Île* this dream, or chimera, has been realized, but probably not in the way Michel envisioned. There is a cult, the *élohimites*, who on the face of it, seems to be an Esalen-like group devoted to self-improvement with a good dose of hedonism on the side. What is not generally known to its members is that most of their money is going into research on immortality of a sort. What makes this novel complex, among other things, is that its protagonist Daniel appears in several reincarnations. The "present body," as the Dalai Lama refers to himself, is Daniel 1, and the final body is Daniel 25, who is separated from Daniel 1 by 23 incarnations and many generations. It is not our intention here to adumbrate how this is supposed to work, and to present much of a literary analysis of the book. Rather the focus is on the science. But, in general, it seems that, to employ a term from calculus, Houellebecq's literary career has reached a point of inflection. He is no longer a "promising" writer. He is over 50 years old, and either he is going to realize his considerable gifts or he isn't. The signs, based on this novel, are disturbing. The gratuitous sex has now proliferated like a fungus. Reading it is like being drilled by a dentist without benefit of anesthesia.

The science in this novel is also in a sense gratuitous. In *Les Particules*, Michel, the principal protagonist, is a scientist. Given Houellebecq's background, it is perfectly reasonable for him to give Michel some real science to do. But, in the latest novel there is no reason to bring in, for example, the quantum theory. I am sure that almost all readers of the novel will skip over this and will not investigate the mathematical jape that Houellebecq tosses in earlier, no doubt for the fun of it. But I know something about these matters and can share what I know.

First, the quantum theory. While the testing of Bell's inequality by Aspect and others was very interesting, in my view it had very little to do with the real dilemma that the theory presents. I think that this was Bell's view also. What the experiment showed was that no local theory with hidden variables could reproduce the results of quantum theory. That is all it showed. There is no evidence that Einstein ever considered such a theory. If I understand him, and I am not sure I do, I think that he wanted to produce a classical theory that would somehow generate the results of the quantum theory. He never really got anywhere with this, so I was never clear exactly what he had in mind, but it was not Bohm's hidden variables. In 1952, Einstein wrote a letter to Max Born in which he asks, "Have you noticed that Bohm believes...that he is able to interpret the quantum theory in deterministic terms? That way seems too cheap to me. But you, of course, can judge this better than I."[124] I think what Bell's work, and what followed, accomplished was to make the subject of the foundations of quantum theory "respectable." Bell's credentials as a physicist were such that physicists were no longer embarrassed to express their misgivings about the foundations of the theory. But I think the attitude of most physicists was well put by the younger of Freeman Dyson's first two children, George, when he was very small. His somewhat older sister Esther, in the way of older sisters, announced one day with considerable authority that she now understood how to row a boat. She said that you pull the "rowers" through the water and create a hole in the water and that is how to row a boat. George replied that he did not understand how to row a boat, but that he could row it anyway. Bell introduced the concept of FAPP—"For All Practical Purposes." Most physicists were quite happy with FAPP, and many still are, but at least it is now acceptable to dig deeper.

Much of this activity has occurred since Houellebecq wrote *Les Particules*. He seems to have followed, at least to some extent, the developments. The profound question to which the final answer, in my view, has not been given, is how does the quantum world, with its probabilities and uncertainties, fit with the classical world which, FAPP, does not show any of this. Bohr insisted that experimental apparatus must be classical. For example, Aspect's experiment, which measures a quantum mechanical effect, must, in the final analysis, do this with an apparatus involving, symbolically speaking, dials and pointers that can be described by classical physics. Otherwise, Bohr would argue, we do not know what we are talking about. The problem, which Bohr never discusses clearly, is where does the quantum world leave off and the classical world begin.

Bohr's vagueness about this was an enormous source of irritation to Bell. Bell had red hair and a red beard and, when he discussed Bohr's "obscurantism," in his lilting Irish accent, he looked positively diabolical. I would place the origin of the present work on the foundations with a relatively short paper that the British physicist P. A. M. Dirac published in 1933 with the rather technical title "The Lagrangian in Quantum Mechanics."[125] In 1930, Dirac published the first edition of his great text on quantum theory, which Einstein declared was the most logically perfect presentation of the theory that he knew. Some decades later, Dirac told an interviewer that the first chapter of his book was missing. This was the chapter where he would have discussed the foundations of the theory. He probably never found a way of doing this that satisfied him. I am sure that, if he had, his 1933 paper would have played a role.

---

[124] *The Born-Einstein Letters*, Walker, New York, 1971, p. 192.

[125] This paper and the subsequent ones I am going to talk about can be found in Feynman's thesis, edited by Laurie M. Brown, World Scientific, Singapore, 2005.

As it was, I do not think too many physicists paid a lot of attention to it until Richard Feynman took it up for his Princeton Ph.D. thesis written under the supervision of the late John Wheeler. Feynman never published his thesis, which was finished in 1942. He spent the war at Los Alamos and after the war, in 1948,[126] he did publish a version of it. But Wheeler was so taken by Feynman's work at the time that he went to see Einstein, thinking that the sheer beauty of the mathematics would be enough to change the old gentleman's mind. He did not understand that Einstein's concerns had nothing to do with mathematics. They had to do with "reality." The late Abraham Pais once told me that on one occasion he and Einstein were walking in the general direction of Einstein's house in Princeton when a full moon rose. Einstein asked Pais if he really believed that the moon was not there when he didn't look at it.

In trying to convey what Feynman did, I am well aware of what gets lost in translation when one attempts to explain ideas involving mathematics without using mathematics. But I will do my best. Let us suppose that at some initial time, t, the state of a system is specified by the measured values of quantities such as position, momentum, energy, and perhaps spin or some other physical attribute. If this system is quantum mechanical, then the uncertainty principles limit what observables can be specified simultaneously. Our specification must reflect that. The system then evolves in time, undergoing interactions with external forces or with forces that different elements of the system exert on each other. At a later time, t', we again make a measurement of all the quantities that specified the initial state. If this was classical physics, the values of these newly measured quantities would be predictable from the initial conditions and the forces that act on the system. This is classical determinism. In quantum mechanics things are entirely different. All the theory allows us to do is to predict the relative probabilities of the various values that the quantities we measure at the later time might have. Until we make such a measurement none of these observables has a definite value. There is only a spectrum of possible values, one of which is selected by its measurement. Wheeler once described this in terms of a quantum mechanical baseball umpire calling balls and strikes. "They ain't nothing until I call them," the umpire insists.

In classical physics the path—an orbit for example—by which the system evolves from one state to another can be traced out. We are certain that there is such a traceable path even if we do not choose to observe it. In quantum mechanics there are also "paths." These are intermediate states through which the system can transit as it goes from its initial to its final state. But, if we do not do observations, all we can do is to innumerate the possible paths and to give the relative probability that the system might follow one path or another. The final probabilities are determined by summing over all the possible intermediate paths weighted by their respective relative probabilities. In fact, things are even more complicated than this because these paths can interfere with each other somewhat in the way that light waves from

---

[126] *Reviews of Modern Physics*, 20, 367–387, 1948. It is reproduced in Brown, op. cit.

different sources can interfere. One can only assign probabilities to these paths if they do not interfere. The term of art that is used is if they are "decoherent." This decoherence of paths will play a vital role in passing to the classical limit of the quantum theory.

When the German physicist Max Planck introduced the quantum of energy into physics at the end of the nineteenth century, he also introduced a new fundamental constant of nature that was associated with these energy quanta. This is now called "Planck's constant," and we still use his original letter "h" to denote it. This constant is very small, which is why we are not aware of quantum phenomena in our daily lives unless we happen to work in a physics laboratory. There are no manifestations of Heisenberg's uncertainty principle that we can detect in the Moon's orbit around Earth, for example. The passage to the classical limit of the quantum theory always involves studying what happens in the theory when you let Planck's constant tend to zero, or at least become very small. In his 1933 paper Dirac has a brief, but very important discussion of this, which Feynman amplifies in his 1948 paper. The basic idea is if you consider the sum over the "paths"—the possible intermediate states that might be transited between the initial and final states—and let Planck's constant tend to zero, then nearly all the paths that might contribute have their contributions washed out in the sum. The only surviving paths are, they argue, the paths that obey the laws of classical physics.

Since Feynman's 1948 paper there has been practically a cottage industry amplifying and clarifying these ideas. I do not intend to review this enterprise. For purposes of this essay I am going to focus on the work of two physicists, Murray Gell-Mann and James Hartle. These are the ones that Houellebecq cites in his novel. Gell-Mann won the 1969 Nobel Prize in Physics for his work on the theory of elementary particles. These are the real elementary particles and not Houellebecq's fictional analogues. Gell-Mann invented and then named the quark. Hartle, who was Gell-Mann's student, is noted for his work on quantum theory applied to the cosmos at large. Some of this work he did with Stephen Hawking. I very much doubt that Houellebecq read any of their technical papers.[127] I know, for reasons that I will explain, that he read Gell-Mann's book, *The Quark and the Jaguar*.[128] In their papers, and in Gell-Mann's book, the term "path" is replaced by the term "history." This does not change anything. To compute the probabilities of outcomes you sum over "histories," histories characterized as "classical" or, as they argue and as I will explain, "quasicassical." A necessary characteristic is that any such candidate of history must de-cohere from the other possible histories. Gell-Mann and Hartle are fond of race-track analogies. The outcome of a given race does not depend on the possible outcomes of the same race that were not realized. But how do the histories de-cohere? What is the physical mechanism? Here the notion of "graining" enters. At any given instant there is vast amount of information available that is just

---

[127] For example, classical equations for quantum systems, *Physical Review* D, 47, Number 8, pp. 3345–3382, 1993. This paper contains a brief review of previous work.

[128] *The Quark and the Jaguar*, by Murray Gell-Mann, W. H. Freeman, New York, 1994.

ignored. Let me take a relevant example. In each cubic centimeter of the universe there are somewhat more than four hundred quanta of radiation—photons—left over from the Big Bang explosion. Unless we are radio astronomers to whom these quanta manifest themselves as background noise in their telescopes, which is how they were discovered in the first place, the existence of these quanta plays no role in the description of any activity we take part in.[129] If, for example, we hit a tennis ball, the fact that this ball collides with this myriad of microwave quanta is irrelevant, FAPP, in predicting the trajectory of the ball. But, Gell-Mann and Hartle would argue, it is these collisions that cause the decoherence of this trajectory. Averaging over all these collisions is what erases the coherence of other candidate trajectories. The term in art for this is "coarse graining." This would be Gell-Mann and Hartle's answer to the question that Einstein asked Pais about the Moon. The Moon is there, and following its classical orbit around Earth, even when you don't observe it, because its collisions with the cosmic photons and other kinds of radiation has made its coarse-grained orbit de-coherent with the other possibilities that quantum theory would allow. One can only imagine how Einstein would have greeted this explanation. However, there are residual effects of this process, which are examined by Gell-Mann and Hartle. The reduction to classical physics is not complete. There are, in general, residual random noise effects. That is why they call the reduction quasiclassical. For the Moon, FAPP, these effects are totally negligible, but more generally they must be taken into account. However, these considerations are not what attracted Houellebecq's attention.

As mentioned earlier, in his novel there is Daniel 1 and what turns out to be his final reincarnate, Daniel 25. Periodically Daniel 25 delivers soliloquies both on his own fate and that of Daniel 1. Daniel 25 has grown restless living in a kind of hermetic colony with the other reincarnates. He leaves to go out into the wider, and, as it turns out, very hostile world, accompanied by the reincarnate of Daniel 1's dog, Fox. In the soliloquy in question he is musing about what "existence" means. He approaches what he calls the paradigm of existence using a construct of Gell-Mann and Hartle. They note that some adaptive systems, such as us, are capable of observing. They give the name IGUS for such a system, an acronym for "information gathering and utilizing system." Such an IGUS, as its history evolves, confronts branches, or paths, that it does not follow. It prunes and discards these unfollowed branches. In his book Gell-Mann writes, "An observation in this context means a kind of pruning of the tree of branching histories. At a particular branching, only one of the branches is preserved (more precisely, on each branch, only that branch is preserved!) The branches that are pruned are thrown away, along with all the parts of the trees that grow out of the branches that are pruned."[130]

It is this paragraph, and especially what is in parentheses that appears to have captured Houellebecq's poetic imagination. He writes, *"Pour un IGUS observateur,*

---

[129] Hartle has pointed out that about 1 percent of the white noise heard on an empty TV channel comes from this background radiation, so in this sense it does enter our "activities"

[130] Gell-Mann, 1994, op. cit., p. 155.

qu'il soit naturel ou artificial, une seule branche d'univers peut être dote d'une existence réelle; si cette conclusion n'exclut nullement la possibilité d'autres branches d'univers, elle en enterdit tout accès à un observateur donné; pour reprondre la formule, assez mystérieuse mais synthetique de Gell-Mann, sur chaque branche, seuele cette branch est préservée."[131]—"For an IGUS observer be it natural or artificial, only one branch of the universe can be given a real existence; if this conclusion does not at all exclude the possibility of other branches of the universe, it forbids any access from them to the given observer; in the formulation, at once mysterious and synthetic, of Gell-Mann 'on each branch only that branch is preserved.'" I am not sure how mysterious this is once one understands the context and the physics. In any event, I asked both Gell-Mann and Hartle, friends of long-standing, if they had discussed this with Houellebecq. Neither one had, and they were quite surprised—and rather pleased—to find themselves in a novel.

The final example I will give involves a mathematical jape. I suppose it could be argued that since Daniel 25 is waxing metaphysical it is not unreasonable to bring in quantum theory, if that is part of his metaphysics. But there is no reason for the jape. It is pure intellectual exuberance, showing off if you want to be unkind. Daniel 1 is a comic monologist both subversive and very successful. He is sort of a French Lenny Bruce without the baggage. He has tired of his work, which made him a multi-millionaire. He decides to take a kind of sabbatical in Spain with his wife. While he was writing this, Houellebecq took a sabbatical in Spain with *his* wife. Daniel 1's wife leaves him, and friends propose as a kind of distraction that he come to a conference of the élohim. Daniel 1 listens for the first time to the "prophet"—the spiritual leader of the sect. In his discourse the prophet asks metaphorically how many will there eventually be of the chosen. In his answer he reels off some of the most significant numbers in the theory of numbers, not because these are the actual numbers of the chosen but as examples. It is this list of numbers that is the jape. I will give one and the story that goes with it, and you will get the idea. The interested reader can find many websites on the others.[132] The number I chose in my illustration is 1,729. The story of this number begins on December 22, 1887, with the birth in the town of Erode in the Tamil Nadu in India, of Srinivasa Ramanujan. At age ten Ramanujan first encountered formal mathematics in school, and, by age 11 he had mastered all the mathematics that was available in the local schools. By 14, he was completing the last of the mathematics examinations that he was required to take in half the allotted time. He had given up studying any other subject. He had come across a book that was used in Britain to prepare students for the very difficult Tripos examination and proceeded to master it. The book contained mainly statements of results rather than proofs. This was the way in which Ramanujan thought mathematics was done. He got a scholarship to the Government College in Kumbakonam in 1904 but lost it because all he was prepared to work on

---

[131] *La Possibiilté*, p. 345.

[132] The list in question can be found in *La Possibilité*, p. 126.

was mathematics. He was never able to get a university education, but he began publishing papers and acquiring at least a local reputation. He finally approached Ramachandra Rao, one of the founders of the Indian Mathematical Society. Rao found him to be wretchedly poor and carrying a tattered notebook that contained some of his results, which Rao could not understand. But he was impressed enough to help him get a job as a postal clerk in 1912, which would earn him enough money to get by on. By this time he had been married for three years to a girl who was nine when he married her by arrangement. In 1912, Ramanujan began sending letters to mathematicians in England containing his results. There were no proofs, simply statements. He had seen a book by the great Cambridge University mathematician G. H. Hardy, and in January of 1913 sent him a ten-page letter stating a variety of results and an explanation of who he was. Hardy was flabbergasted by what Ramanujan had sent him. One can find some of these results on the web. When you see them you can't imagine how anyone could have invented them.

Since there were no proofs in many cases, Hardy couldn't see them, either. But he realized that he was in the presence of a true mathematical genius. He immediately wrote to Ramanujan to ask if there was anything he could do. Ramanujan replied that Hardy might help him to get food since he was starving. Hardy helped to get Ramanujan a scholarship at Madras University and then decided to bring him to England. This posed a problem because Ramanujan was a Brahmin and was afraid of losing his caste if he made a foreign voyage. Fortunately his mother had a propitious dream, and in 1914 Ramanujan came to Cambridge.

There then followed four years of extraordinary collaboration with Hardy. In addition to their published work, Ramanujan filled notebooks with outpourings of novel and profound mathematics, which mathematicians are still studying. Now to the number.

This number has become famous to mathematicians because of an anecdote involving Hardy and Ramanujan. The only version of it that I know was related in a monograph on Hardy by C. P. Snow. Snow's version has obvious errors, which makes me wonder. But the anecdote is now part of the folklore, like Newton and the apple. Ramanujan had always been in poor health. This is not surprising, considering what must have been his chronic malnourishment. During the war in England he could not get enough of the fresh vegetables he needed for his vegetarian diet, and he again became ill. In 1917 he was hospitalized. Snow informs us that Hardy came to visit what Snow says was the dying Ramanujan. Ramanujan was not dying and, in fact, returned to India in 1919. There, his health failed again, and that year he did die, of amoebic hepatitis it seems, which he had contracted earlier in India and which would have been curable with the then available drug emetin[133] if only Hardy had taken him to see an expert in tropical medicine. In any event, as the

---

[133]I am grateful to Freeman Dyson for calling my attention to this and also pointing out that Ramanujan often went to the Sarangapani temple in Kumbakonam and did mathematics during the service. There is etched on the wall the number 2719, which has an even deeper mathematical origin.

story goes, Hardy came to visit him in the hospital and said that he had ridden in a taxi with the number 1,729, which he thought was a rather dull number. Ramanujan immediately replied that in fact it was a very interesting number as it was the smallest number that could be expressed as the sum of two cubes in two ways. That Ramanujan would know this does not surprise me, but that Hardy did not does surprise me. In 1657, one Bernard Frénicle de Bessy published the fact that $1,729=1^3+12^3$ and $1,729=9^3+10^3$. Try it out on your pocket calculator. Mathematicians call this the "taxicab problem," and, in their lingo 1,729 is known as Taxicab(2). They have studied what the smallest number is that can be expressed as the sum of two positive cubes in three ways, Taxicab(3), and so on. Mathematicians are like that.[134]

Do you need to know any of this to read Houellebecq's novel? Of course not. However, it is part of the creative imagination that informs the novel. We accept the fact that novelists use what they know and have experienced consciously or unconsciously when they write. For Houellebecq part of that experience was science.

---

[134]Taxicab(1) is $2 = 1^3 + 1^3$. Taxicab(3) = 87539319
  $=167^3+436^3$
  $=228^3+423^3$

Chapter 15
Topology

Ernest Solvay was a wealthy Belgian industrialist who dabbled in physics. He got the idea of sponsoring a conference to which the most notable scientists of the time would be invited with the hope that they might spend some time discussing his ideas. The first of these meetings was held in Brussels in 1911, and there is no record that I am aware of that anyone discussed Solvay's ideas. The guest list could be taken from the authors' index of any text on twentieth-century physics. Madame Curie was there, and just behind her in the photograph, and standing to Einstein's left, is Paul Langevin, a theoretical physicist with whom she had just ended a notorious affair. Seated next to her on her left is Henri Poincaré, who was one of the greatest mathematicians who ever lived. But he was also a polymath who had worked in all sorts of sciences, including physics. Some of the things he wrote before Einstein's invention of the theory of relativity in 1905 have intimations of the theory and, after its invention, he developed some of its mathematics.

Poincaré died at age 58 in 1912, the year after the conference. As far as I know the conference was the only time the two men met, and Einstein was not impressed. On November 15, 1911, Einstein wrote to his Swiss friend and confidante the physician Heinrich Zangger, "*Poincaré war (gegen die Relativitätstheorie) einfach allgemein ablehnend zeitge bei allem Scharfsinn wenig Verständnis für die Situation.*"—Poincaré was simply generally antipathetic (in regard to relativity theory) and showed little understanding for the situation despite all his sharp wit.[135] Fortunately for Einstein, the feeling was not mutual. Shortly after the conference Poincaré wrote a letter of recommendation for Einstein that began, "Monsieur Einstein is one of the most original minds that I have known; in spite of his youth [Einstein was then thirty-one] he already occupies a very honorable position among the leading scholars of his time."[136] Einstein got the job, which was at his alma mater, the ETH, in Zurich.

---

[135] This can be found in *Subtle is the Lord* by A. Pais, Oxford University Press, New York, 1982, p. 170. I have taken Pais's English translation.

[136] Pais, op. cit., p. 170.

When Poincaré died his funeral was a national event. In his book *Poincaré's Prize*[137] the mathematician and journalist George G. Szpiro explains why. Poincaré was born into a middle class intellectual home in Nancy. His father was a doctor. Because of an illness, young Poincaré was tutored at home until the age of eight. Once he got to school he was first in everything, although he did not focus on mathematics until he reached fourteen—somewhat late for mathematicians, who usually show proclivities almost before they can speak. As I mentioned in connection with Houellebecq, the French have a system—*les grandes écoles*—which while often derided, especially by people who were not able to get into one—is a wonderful institution for discovering and nurturing gifted young people who later become, in many fields, the elite of France. In 1873 Poincaré was accepted at one of them, the *École Polytechnique*, the "L'X," which was located in Paris near the Sorbonne. It was, and is, a military school to which women were admitted starting only in 1972. Its mission is to train engineers, some of whom would serve in the military.

I spent the fall and winter of 1959–1960 there working with the late Louis Michel—a noted French theoretical physicist and an alumnus. It was still located in the Latin Quarter. Now it has moved to a campus outside Paris. There was always a very strong tradition of mathematics at the school, which was certainly true when I was there. Poincaré was a brilliant student in everything but drawing, at which he was completely inept, so that he only finished second in his class. After graduation he began his service in the *école des Mines*, where he trained as a mining engineer. The polytechnicians all did service of this kind. Michel, for example, was in the *Poudres*, which, theoretically, occupied itself with gun powder explosives in general. When Poincaré graduated in 1878 he became a mining engineer, something that he took very seriously. In 1879 he investigated a tragic accident in a coal mine and wrote a report that is a model of its kind. But he had also begun to publish papers in mathematics, and these were so impressive that the very same year he accepted a professorship in mathematics at the University of Caen.

Throughout his career Poincaré was notorious for writing papers that had loose ends. Some of these were relatively simple to fix, and some were not. Among the latter was a paper that Poincaré wrote for a mathematics prize contest sponsored by King Oscar of Sweden. Poincaré won with an essay on showing that three bodies acting under their mutual gravitational attraction, if they don't collide, follow trajectories that can be predicted into the indefinite future. This is a notoriously difficult problem, and, after Poincaré had been declared the winner, it was discovered that his paper was wrong, or at least incomplete. This generated a considerable commotion, and in repairing the damage Poincaré created the beginnings of what has become modern chaos theory. He also invented a discipline that he called *analysis situs*, now commonly known as "algebraic topology," which will shortly bring us to Poincaré's conjecture, which is the subject of both Szpiro's book and that of

---

[137] Dutton, New York, 2007.

Dona O'shea, who is a professor of mathematics at Mount Holyoke College.[138] In 1881 Poincaré was appointed a professor at the University of Paris, where he remained for the rest of his life. In 1909, he was elected to the Académie Françse. Somewhat earlier he had helped to defend Alfred Dreyfus, a French army captain and also a *Polytechnique* graduate, who had wrongfully been accused of treason. Poincaré refuted the "scientific" arguments against Dreyfus. Poincaré was a brilliant stylist. I remember reading his *Science and Hypothesis* as an undergraduate long before I had any idea of what he had done in mathematics. It was like taking a drink from a clear stream.

Before we get into Poincaré's conjecture I must give fair warning to the reader. It is true that I have a master's degree in mathematics. It is also true that I once graded papers while at Harvard for one of the most noted topologists of his day, Hassler Whitney, who makes a cameo appearance in Szpiro's book as the creator of "Whitney's trick," which is not explained and in O'shea's as being in the 1930s at Princeton. Not withstanding, when it comes to the subject matter of these books I am a layman. In this respect I would have been vastly helped if Szpiro had drawn a few diagrams. There are none. When I read something like this: "Take a dodecahedron, the solid that is bounded by twelve regular pentagons. Two pentagons always lie on opposite sides of the dodecahedron, so there are six pairs. Identify one pair, stretch and bend the dodecahedron so that-after twisting it by one fifth of a revolution...etc." my mind freezes. However, keeping in mind Bohr's admonition that one should not speak more clearly than one thinks, I will try to explain the conjecture.

Imagine that you are given a blackboard and as much chalk as you want and asked to draw all possible distinct closed figures—triangles are an example—on the blackboard. Clearly this request as stated is absurd. You will be at it for as long as you live, and could be at it until the end of the universe. To make this into a sensible request we must specify what we mean by "distinct," or at least what topologists mean. To a sensible layperson, the drawing of a triangle and a circle seem distinct. But a topologist would say that you could deform the triangle into the circle and vice versa without tearing anything so that from a topological point of view they are identical. Now we are getting somewhere.

In two dimensions a rough rendering of the Poincaré conjecture is that if you draw figures on the two-dimensional surface of an object embedded in ordinary three-dimensional space, and if all these figures can be shrunk to a point—no holes—then topologically speaking this is the surface of a sphere. For two dimensions it was not, at the time of Poincaré, a conjecture. It had been proven. What he did was to propose in 1903 that there was a three-dimensional version. This is hard—perhaps even impossible—to visualize, since it is a three-dimensional object embedded in a four-dimensional space. He never could prove this, and neither could anyone else for a hundred years, until it was finally proven in 2003 by a rather

---

[138] *Poincaré's Conjecture*, Walker, New York, 2007.

curious Russian mathematician by the name of Grigori "Grisha" Perelman. Oppenheimer once said that Wolfgang Pauli was the only person he knew who was identical to his own caricature. I have known a number of mathematicians, including several who are mentioned in this book. You can trust me that many of them are identical to their caricatures. What I greatly enjoyed about Szpiro's book is the affectionate way he describes these people, including Perelman. His account in this respect is more savorous than O'shea's.

Perelman was born in 1966 in what was then called Leningrad (now St. Petersburg) to a Jewish family. His father was an electrical engineer and his mother a teacher of mathematics. If I ever have the opportunity to interview Perelman I will ask him what his first mathematical memories were. As I mentioned before, real mathematical ability often declares itself early. Here are two examples I know of. Freeman Dyson once told me that when he was still young enough to be "put down for naps" he invented what he later learned was the notion of the convergent infinite series. He began adding $1 + 1/2 + 1/4 + 1/8\ldots$ and noticed that the sum was approaching 2. His teacher Hans Bethe told me that when *he* was very young he discovered that if you had a very big number and took its reciprocal, it became a very small number—something that fascinated him.

In any event Perelman was enrolled in a Leningrad school for children who were especially gifted in mathematics. In 1982, he took part as a member of the Russian team in a Mathematical Olympiad for high school students. He got a gold medal, having achieved a perfect score on the problems. After this he enrolled in the Leningrad State University, from which he took his Ph.D. He then joined the St. Petersburg branch of the Steklov Institute of Mathematics, which was run by the Russian Academy of Sciences. This was followed by several appointments in the United States at places such as the University of California at Berkeley. All of this is mentioned for a reason. Einstein used to say that the ideal job of a theoretical physicist was as a lighthouse keeper. Then you could think all day with no one bothering you. The downside is that you would not have the slightest idea what to think about. Even an Einstein needs a context. Years ago, when I used to trek in Nepal before it was popular, I would meet one or two people in remote Sherpa villages who were regarded by their neighbors as geniuses. These people were, at the time, almost completely disconnected to the outside world, which resulted in their genius being devoted to their religion. They were monks or shamans. While such a person might come up with some interesting ideas involving numbers, when it came to something like the Poincaré conjecture or the Riemann Hypothesis—another deep mathematical mystery—they would be clueless. Each of these areas of mathematics has a century of scholarship behind them, and unless one had mastered this, one would have gotten nowhere. Even stating the Riemann Hypothesis, for example, requires some mathematical sophistication.

Perelman did not do his early work in isolation but rather at some first rate mathematics institutes. As good as this work was, no one, as far as I know, anticipated the three papers in 2002–2003 which he published on the Internet in what is known as "arXiv."

The term "arXiv," pronounced "archive" (the "X" is really the Greek letter "X," or chi) is the brainchild of the theoretical physicist Paul Ginsparg.[139] When Ginsparg came up with it in 1991, he was at the Los Alamos National Laboratory. He is now at Cornell. The internet was just getting started, and it occurred to Ginsparg that the time-honored method of physicists sending out paper preprints was soon going to disappear. What would be needed would be an electronic archive in which papers could be stored and accessed by anyone who was interested.

Ginsparg, who, among other things, is a computer wizard, put this together. At first it confined itself to physics but, by now, it covers several fields, including mathematics. It is very easy to use. You browse the site, pick a field, and then an author. Press a button and there you are. In the beginning, essentially any paper that resembled physics could be archived. There was no refereeing system. There still isn't, but to get a paper onto it one must have an endorser, someone who presumably has had a paper archived. Recently Ginsparg came up with a new wrinkle involving plagiarism. Papers will be scanned to see if they too closely resemble any of the other 400,000 in the archive. The author(s) will have this pointed out to them and the overlap spelled out. Ginsparg may have envisioned a future in which the usual refereed print journals would disappear and everything would become electronic. This has not happened, at least not yet. Scientists, by and large, still want to be published eventually in a peer-reviewed print journal of record. Most of them use arXiv as a good place to establish priorities and perhaps get comments. To this, and to nearly everything else, Perelman is an exception. The *only* place he ever published his three papers was on arXiv. His attitude was that anyone who wanted to learn what he had done could learn about it there.

To someone who has not followed the evolution of the work on the Poincaré conjecture over the last century, the titles of Perelman's three papers would appear completely baffling. They all have to do with what is called "Ricci flow." For example, one of them is called "Ricci flow with surgery on three-manifolds." And in none of the titles does the term "Poincaré conjecture" appear. Most of these books deal with how the Poincaré conjecture morphed into a proposition about "Ricci flows"—indeed what they are—and about the mathematicians responsible. It is largely a story of valiant attempts by mathematicians who were obsessed, but could never quite get there. This persisted more or less until the early 1960s, when a young mathematician named Steve Smale proved the conjecture for all dimensions greater than five. Poincaré only conjectured something about the three-dimensions. In this case it turned out to be easier to establish his conjecture in a higher dimensional space, one in which we do not live, but in which there are analogous geometric figures. We will not explain here in any detail why this is so. In any event, in 1981, in what was regarded as a mathematical tour de force, a young mathematician named Michael Freedman proved it for four dimensions. This left three dimensions, the original Poincaré conjecture.

To give at least a flavor of how this problem had evolved by the time that Perelman got into the act, let me say something about the two-dimensional situation.

---

[139] I am very grateful to Paul Ginsparg for comments on this essay.

The surface of a three-dimensional object is two dimensional. If you were a flat bug crawling along this surface you would not know that you were on the surface of a three-dimensional object. Viewed this way, it turns out that there are only three topologically primitive two-dimensional objects. They can be made into surfaces of spheres or tori or even more complex objects. A planar surface can be rolled up and re-shaped so that it looks like a bagel. This surface has one hole in it so it is called a surface of "genus one." By suitable contortions one of the primitive objects can be manipulated so that it morphs into an object with two holes or more. Such an object has genus two or higher. The surface of the sphere has genus zero. Any curve on the surface of a sphere can, by continuous deformations, be shrunk to a point. This is not true of the surface of a bagel, for example.

The Poincaré conjecture in two dimensions is that any surface in which all the curves can be shrunk to a point continuously is topologically equivalent to the surface of a sphere. As mentioned earlier, at the time of Poincaré this was not a conjecture but a theorem.

For three dimensions things are vastly more difficult. In the early 1980s a mathematician named William Thurston made a conjecture, which he could only partially prove, that in three dimensions there were eight primitive topological objects of which the sphere was one. This was a deeper conjecture than the one made by Poincaré, since the Poincaré only said that a certain class of three-dimensional objects was topologically equivalent to the sphere, while Thurston was claiming that any three dimensional object fell into one of only eight categories. A little later the mathematician Richard Hamilton changed the paradigm. He did this by making use of some work that the Italian Gregorio Ricci-Curbastro had done at the end of the nineteenth century. Ricci's work had nothing to do with the Poincaré conjecture per se. It rather had to do with the geometry of curved spaces.

To see what is involved, let us go back to our flat bug that is crawling around the surface of some object. How could the bug tell whether this surface was curved or flat? If we are dealing with a very sophisticated bug it would draw a triangle. It would then measure the interior angles of same. It would recall a theorem from plane geometry that the interior angles of a triangle drawn on a flat plane add up to 180 degrees. But suppose the bug was on the surface of a sphere and drew a triangle. Here we should be a little careful about what we mean by a triangle. What replaces the straight lines on the plane are the arcs of circles—great circles—whose centers are the centers of the sphere. Earth's equator is an example. Now suppose we make a triangle whose base is a piece of the equator and whose sides are great circles that end up at the North Pole. If you draw such a circle you will see that the base angles are 90 degrees but that there is an excess angle at the top. The angle sum is *greater* than 180 degrees. When this happens one says the curvature of the surface is positive. If you carry out a similar process on a saddle-shaped surface you will find that the angles add up to *less* than 180 degrees. In this case one says that the curvature is negative. What Ricci did was to find a characterization of the curvature of a surface when it varies from point to point. This analysis was largely of interest only to mathematicians until Einstein used it in his theory of general relativity and gravitation, which he produced in 1915.

Einstein realized that gravitational fields bend light rays. Thus a light ray from a distant star that grazes the Sun will be bent slightly, causing an apparent displacement of the star as compared to its position when the Sun is elsewhere. This was confirmed in the famous solar eclipse expeditions of 1919. One could imagine making a triangle out of three light rays and then measuring its angle sum. If the light was in the presence of gravitation the triangle would not be Euclidean. Actually the situation is more complex because gravity also affects clocks. Thus the geometry involved is four-dimensional with the fourth dimension being time. The Ricci analysis is readily extended to higher dimensions. What Hamilton did was to imagine a three-dimensional "manifold"—a surface-that is all bunched up. He presented an equation that describes how this manifold evolves toward one in which the curvature is smoothed out. This evolution is what he called the "Ricci flow." Hamilton was able to show that if the manifold has positive curvature everywhere, like the surface of a sphere, then basically it is a sphere. The surface will, in the course of its evolution, collapse into a point. This is a special case of the Poincaré conjecture. But if negative curvatures come into play then there are potential disasters. There can be places where the equations blow up—singularities—so if these singularities cannot be tamed then no conclusions can be drawn. This is what Perelman did. He showed that Hamilton's Ricci flow method can be made to work for any three-dimensional manifold and that the evolution will be into one of the eight primal shapes conjectured by Thurston. This not only proved the Poincaré conjecture but Thurston's as well. It was a monumental and astonishing piece of work.

Perelman posted the first of his papers on November 11, 2002, and the last on July 17, 2003. I have looked at the first one. One of the things that struck me was the fluidity and clarity of the English. There is no hint that it was not written by someone whose mother tongue was not English. I was also struck by the no-nonsense nature of the paper. Over the years I have been exposed to a number of crank papers in theoretical physics. One hallmark of a crank is usually the shrill tone of the paper. The crank has had an epiphany, and you are going to share it whether you like it or not. In Perelman's paper there are no claims apart from the theorems that are proved one after the other. It is up to the reader to understand the dramatic consequences of Perelman's analysis. Whatever else, this is not the paper of a crank. It did not take long for mathematicians to get the gist. Perelman was invited to give various lectures in the United States. At some of them correctible flaws in his arguments were noted to which he was very responsive. In every way he appeared to be functioning like a normal mathematician, but then there was a series of events that seem to have led to his present self-imposed isolation and possible abandonment of mathematics.

I do not know Perelman, so I am not sure what led to what. Several groups of mathematicians dedicated themselves to understanding and explaining Perelman's results and methods. Certainly no one, including Perelman, could object to this. However, some of the groups went somewhat overboard and confused explanation with discovery. Because they were able to explain Perelman's work more clearly than Perelman had chosen to do, they decided that they must have discovered it. This led to wrangles and even threatened lawsuits that still seem to be going on. Perelman had no involvement with any of this and simply withdrew. Among other things he refused the 2006 Fields Medal, which is the most prestigious award any mathematician not over 40 can receive. It is often referred to as the mathematician's Nobel Prize. Prior to Perelman no one had ever turned it down. He has also shown no interest in the million dollars that a wealthy donor named T. Clay has offered by way of a prize. Finally, he has also resigned his position in the Steklov Institute. It is not known how he earns his living or spends his time. For a while he did answer e-mails from mathematicians who wanted clarification, but this has stopped. One can only wish him well. He has already accomplished more than most people manage in a lifetime.

Chapter 16
What the #$*!?

In the late 1970s a book appeared that was called *The Tao of Physics*. It was written by a theoretical physicist named Fritjof Capra, whose intent was to show parallels between some ideas of elementary particle physics and Eastern mysticism. When I heard about it, I decided that I would not read it. I had by this time taken several trips to Nepal, where I had been exposed to the real thing, and I felt that I did not have to deal with some second-hand imported version. But eventually I changed my mind. As a physicist I found myself being asked questions concerning it, so I thought that I should have some idea what it was. Furthermore, I was then writing a biannual column for *The American Scholar*. I thought that if this book was as bad as I expected to be, it would be grist for a column. My expectations were realized on both counts, and I wrote a column that I called "A Cosmic Flow." In it I pointed out that not only were Capra's parallels absurd, but the physics on which they were based was completely outdated. In fact, I noted, that anyone who bases their religious beliefs on some specific scientific model is asking for trouble. It is like trying to build a house in the tropics whose foundation is a block of ice.

I now regret that this foray occurred at a time before my late friend John Bell had met the Dalai Lama and discussed physics with him. The Dalai Lama came to visit CERN in Geneva, where Bell worked, and Bell was selected as one of the delegation to meet him. Bell told me that the Dalai Lama could see no connection between what he understood of quantum theory and Tibetan Buddhism.

What intrigued the Dalai Lama was that so much of the atom seems to be empty space. Why does the whole thing not collapse? Bell was able to explain to him that this is one of those things quantum mechanics accounts for with the stable electron orbits that the theory predicts. By this time Capra had been heard from. He wrote a rejoinder to my column in which he said that my problem was a feeling of sexual inadequacy. I always wondered which of my ex-girlfriends he had talked to.

I had forgotten all about this when not long ago people started asking me about a new film called *What the #$*! Do We Know?* My first thought was that the Great Wheel had turned and that Capra had been reincarnated in a new guise. Once again I swore that I would not see it, but once again after all the questions I gave in.

It is a difficult film to describe, in part because it does not make any sense. A young woman named Amanda appears to be undergoing a particularly tedious form of emotional crisis, she is played by Marlee Matlin, the deaf actress who was such a marvel in "Children of a Lesser God." Since that film she has put on a good deal of weight, which in the present film she rants about constantly while dressed in her underwear. She also pops pills which, unfortunately, she does not share with the audience. They might have helped.

In the film Amanda is a photographer assigned to photograph a Polish wedding. The people look as if they had stepped out of a Diane Arbus collection of freaks. She also attempts to play basketball with a young black boy who looks as if he has accidentally wandered in from another movie, or another life. While all of this is going on, the narrative, such as it is, is interrupted by various obscure pronouncements on quantum theory by individuals who are not identified until the end of the film. I think that I may have had a few of them in class. One of them that I certainly did not have in class was one J. Z. Knight, whose present body is being channeled through Ramtha, or perhaps it is vice versa. It is hard to tell. Ramtha is a Brunhildesh-looking blonde who at one point discusses erections as an illustration of the mind-body experience. One wonders whose.

Most of the discussion of quantum theory is so confused that it falls into Pauli's category of not even being wrong. But there is one thing that is specifically wrong and is such a common misunderstanding (Tom Stoppard, one of my favorite playwrights, wrote a whole play called *Hapgood* based on it) that it is worth while trying to sort it out.

In quantum mechanics it is very important not to make statements about phenomena that one has not observed. A prime example is given by the two-slits. There is an obstruction that blocks the passage of particles such as electrons. However the two slits then can be opened and closed. This allows the passage of electrons though one or both slits. If electrons were marbles, then if only one of the slits was opened, they would presumably follow the same trajectory and all hit some spot on a detector. But electrons are not marbles. They have a wave nature. If one of the slits is left open the electrons will produce a characteristic pattern of maximal and minimal intensities at the detector that reflects this. The same sort of thing will happen if we close the first slit and open the second. What happens if we leave both slits open?

In this case the two slits act like a diffraction grating. The waves going through the slits interfere, and this produces a pattern of maxima and minima at the detector. Once you accept the wave nature of the electron you will probably find this not too hard to swallow. It is how the wave nature of light was first demonstrated. But now I am going to do something a bit different. I am going to leave both slits open but send the electrons through one after the other. The remarkable thing is that they will still eventually reconstitute the diffraction pattern. This is the point where people begin to go off the rails. If they have not really understood the theory they will say that what this means is that the single electron must have gone through both slits and is therefore in two places at once. Stoppard's play is about spies, and this quantum effect seems like a handy thing for a spy to have.

But before we get carried away we have to ask how do we know which slit the electron has gone through? The only way we can tell is to close one of the slits.

If we do this the diffraction pattern disappears. Hence, with the two slits open, to ask which of the slits the electron went through is a meaningless question. The electron is not in two places at once. It is in no place until you do an experiment that measures its position. In quantum mechanics you must always ask "How the #$*! do you know?"

By the way, the same reasoning applies to Schrödinger's famous cat. This unfortunate animal is locked in a box with a cyanide source that is activated by the random decay of a radioactive element. If we, for some obscure reason, decide to describe this situation quantum mechanically we would say that, in the absence of observation, the cat is characterized by a wave function that indicates a fifty-fifty chance of the cat's being alive or dead. Some people have argued that this means that the cat is both alive and dead—the electron is in two places at once. But you must open the box to find out, at which time the cat will either be alive or dead. If you don't open the box you cannot say. I once visited Schrödinger in his apartment in Vienna. There was no cat and I was told that he did not even like cats. I wonder what he would have had to say about this movie.

Author's Note About "Beating the System"

In the beginning of this book, we discussed economics. Now, here at the end, we will return to it. We have come full circle. It is remarkable that the ideas of physics play such a big role in economics. When Einstein was studying the diffusion of tiny particles in liquids he was able to explain their erratic motion – the Brownian motion – by a model in which these particles were subject to the incessant and erratic bombardment by the molecules of the liquid. This produces the random walk of these particles. The first essay dealt with the application of these ideas to the evolution of the prices of stocks. But Einstein also recognized that there would be fluctuations. Every once in a while the drunkard, who lurches with equal probability in all directions, will make several lurches in the same direction. This can also happen with the stock market, and this essay deals with beating the system by using these fluctuations.

Broadly speaking there are two theories as to how the stock market works, each one vehemently supported by its own cadre of pundits. On the one hand there is what is known as the "efficient market hypothesis." This comes in various forms, but basically it proposes that, relative to the information generally available, all stocks at any given time are fairly priced. You can't beat the system. You might as well pick stocks by throwing darts or by investing in an index fund. The future of the market is determined by a random walk. On the other hand, there is an equally distinguished group of pundits who argue that you *can* beat the system, basically because you can take advantage of fluctuations. These people also believe in an efficient market in the sense that they assume that these fluctuations relax-dissipate-back into the norm. In the first essay we saw how Bachelier viewed the market in terms of Brownian motion. In Brownian motion there are also fluctuations whose dissipation can be described mathematically.

Here is an example taken from a lecture given by Edward O. Thorp, whom we shall meet in the essay to which this is the prologue. Suffice it to say that Thorp is one of the most successful money managers who has ever practiced that craft. In his lecture Thorp notes that, in the year 2000, the company 3Com decided to spin off its Palm Pilot division. This news was widely reported in places such as *The New York Times*. On March 2, 2000, 6 percent of Palm Pilot was offered to the public as an Initial Public Offering. Moreover, the value of these shares was $53.4 billion while the value of the rest of 3Com, which included the 94 percent of the shares of Palm Pilot that were not in the initial offering, was only $28 billion. Within six months the company intended to distribute these remaining Palm shares to its 3Com stock holders. To someone like Thorp the implications of this were evident. You could borrow money to buy as many of the Palm Pilot shares as you could get your hands on. You could then sell them at once at the higher price. You would then wait until the rest of the Palm shares become available at a lower price and buy enough to cover your loan and pocket the difference. Of course, if the Palm shares went up in the interval, your investment would be toast, since you were obligated to return the borrowed shares in a predetermined time period in any event. Here you relied on the efficient market to dampen the fluctuation.

It is interesting that both Bachelier and Thorp, although not Einstein, got their inspirations from the theory of gambling. In Thorp's case it was blackjack. If you read on you will see why there *is* a theory of blackjack and how it might suggest these wider applications.

Beating the System

"Markets can remain irrational longer than you can remain solvent."
– John Maynard Keynes

Up until August of 1957, I do not remember ever having played a card game for money. This had nothing to do with morals. I had been an undergraduate mathematics major at Harvard, going as far as getting a Master's degree before switching over to theoretical physics. During this whole time I had been doing mathematics day and night and, playing a card game like bridge, or even poker, where some mathematical skills were involved, was the last thing I wanted to do. Better to go to a movie. In the spring of 1957, when I was completing a two-year post doctoral at Harvard, I was approached by a recruiter from Los Alamos to go there as a summer intern. Many members of the Harvard physics department had been to Los Alamos during the war, and one of them must have recommended me. The weapons laboratories such as Los Alamos and Livermore, in light of the then raging Cold War, were expanding and actively looking for qualified people. I had no particular interest in working on nuclear weapons, but I did have a great curiosity about Los Alamos, which was heightened by the fact that I was about to go to the Institute for Advanced Study in Princeton, of which Robert Oppenheimer, whom I had met, was the director. So I accepted the Los Alamos offer, which was conditional on my being able to get a so-called 'Q-clearance' – a very rigorous clearance needed by anyone working in the technical divisions of the laboratory.

Soon after I got to the laboratory it became clear to me that I was going to have no assignment and would have nothing whatever to do with the weapons that were being designed there. It was also made clear that while I had a Q-clearance, information was only shared on a "need to know" basis. Since I did not need to know anything, I was not told anything. I had a colleague, also a post doctoral from Harvard, in the same situation. We shared an office, and I proposed that we work on a problem in elementary particle physics that I had thought of but lacked the mathematical skills to do myself. During most of that summer we happily worked on our problem while, no doubt, bomb designing was going on all around us. We wrote up our paper, and the head of the Theoretical Division, Carson Mark, encouraged us to publish it and identify the work as having been done at Los Alamos. He wanted the image of the laboratory to be something other than a bomb factory.

During the summer I made friends with a more senior physicist named Francis Low. He had just been made a professor at MIT and was spending the summer in Los Alamos with his family. Francis also was not working on weapons, so I was surprised when, in the middle of August, he announced that he was going to Mercury, Nevada, to see some tests. In fact, my first thought was, "I want to go, too." Francis suggested that I ask Carson, which I did. He was agreeable if I paid my way on commercial airlines and the like, which I was more than willing to do.

It turned out that in 1956, four men, Roger Baldwin, Wilbert Cantey, Herbert Maisel, and James McDermott, had published a paper in the *Journal of the American Statistical Association*,[1] called "The Optimum Strategy in Blackjack, "which was followed in the fall of 1957 by their 92-page monograph *Playing Blackjack to Win: A New Strategy for the Game of 21.*[2] I will come back shortly to how they happened to write such a paper and book and, indeed, why there is such a strategy, after I describe the impact of this paper at Los Alamos.

Mercury, Nevada, where the tests took place, is about 65 miles northwest of Las Vegas. As far as I could later determine, everyone connected with these tests, from the scientists to the soldiers on maneuvers, went to Las Vegas to play blackjack. Until 1956, there were probably as many strategies – essentially all losing – as there were players. But Baldwin et al. had found the Rosetta stone. It turned out that they had done their calculations on primitive electro-mechanical hand calculators. But Los Alamos had what was probably at that time the most powerful electronic computer in the world – the so-called MANIAC – used primarily to design weapons. It was elementary to program this computer to run blackjack hands by the tens of thousands to verify that the scheme actually worked. (It was also programmed to play a primitive game of chess without bishops and succeeded in beating a novice.) I have read some accounts of this blackjack history with nary a mention of Los Alamos. I suspect that this activity at the laboratory was not widely known.

The net result of this was that a little pocket folder was produced from which one could readily read off the basic strategy. I use the term "basic strategy" deliberately. It became the basis of the more advanced strategies that included it. Francis had taken the trouble to read the original article and noticed that if you followed the basic strategy you broke about even with the casino. One may recall A. J. Liebling's gambler's prayer: "Dear Lord, help me to break even, because I sure need the money." In fact, the original authors said that the casino would come out marginally ahead, while later calculations using electronic computers showed that the player

---

[1] Vol. 51,429–439

[2] M. Barrows & Co., New York.1957. The book has long been out of print but is being re-issued.

would come out slightly ahead. On the other hand, if you were prepared to do more work and follow a more advanced strategy you could beat the casino by a couple of percent. Francis made an estimate that ran something like this: Suppose I can play a new game every three minutes – a not unreasonable assumption for casino black-jack – and suppose I play for an hour. That is twenty games. Suppose that on the average I bet $10 a game. Part of the advanced strategy is varying the size of the bets, and in those days a $10 maximum was substantial. Then I stand to make about $2 an hour if my advantage is one percent. This is a lot less than the dealer was being paid, but in principle one had the satisfaction of beating the house.

In 1953 the above-mentioned Baldwin, who had a Master's degree in mathemat-ics from Columbia, was a private in the army. He had been drafted and because of his technical background was assigned to the Aberdeen Proving Ground, not far from Baltimore, which did research on ballistics and the like. He liked playing cards, and one of the games he played the day of the revelation was "dealer's choice." The dealer chose blackjack, and Baldwin did not know the rules. In partic-ular he had not realized that in casino blackjack the dealer is an automaton. The dealer makes no choices but simply follows the house rules. To understand this let me explain a bit of how blackjack is played in a typical American casino. (The British have somewhat different rules, as do the French. Even in the United States rules may vary from casino to casino.)

The players, from one to seven, sit around a table with the dealer at the head of the table. I will consider the case where a single 52-card deck is used. In a modern casino several decks are used, which increases the advantage to the house. The dealer shuffles the deck, and one of the players cuts the cards. A single card is "burned," placed face up at the bottom of the deck. In the basic strategy, where you do not count the cards that have been played, it does not matter whether this card is made visible or not. In the more advanced strategies it does matte slightly, and one wants to know what this card is. In any event, the burned card is not played, so effectively the deck has 51 cards. One card matters, but not much. Better not to make a fuss and get unwanted future attention. The dealer then proceeds clockwise around the table, distributing two cards to each of the players. In many casinos these cards are distributed face down. Face up is common now because that keeps the players' hands off the cards and eliminates some forms of player cheating. In the recent blackjack-related film *21*, they were distributed face up, which enhanced the visual effect at the cost of some verisimilitude. (This was a small loss of verisi-militude compared to the others in the film.) The dealer also receives two cards, one of them face up. Now the play begins.

Each player plays in turn before the dealer plays his or her hand. This is what gives the casino its advantage. The object of each player is to have cards that add up to a higher number than the dealer's, provided that the total is 21 or less. On the first deal, with the ace counting by choice as 1 or 11, and all face cards counting as 10, no one can have a higher number than 21. Being dealt a 21 – a blackjack – wins unless the dealer also has blackjack, which is a "push" with no money exchanged. (Incidentally the name "blackjack" comes from a now defunct nineteenth-century practice in American casinos of paying a high premium if the player was dealt the ace of spades and a jack of spades or clubs.) A blackjack pays off 1.5 times your original bet. When it comes your turn, you can ask for an additional card – a "hit" – but, if after whatever hits you take, you go over 21, you go "bust" and lose without the dealer having to beat you. This is where the asymmetry between the players and the dealer resides. Once everyone else has played their hand the dealer plays his or hers. It is here where the dealer becomes an automaton. If the total is 16 or less the dealer *must* take hits until the total is 17 or more, at which point the dealer "stands." There are some nuances involving aces, but this is the general idea. If the dealer goes bust then all the surviving players win.

This was explained to Baldwin, who immediately realized that there must be an optimal strategy. He must also have realized that mimicking the dealer will not work. It turns out that if you do this, the asymmetry between the order in which the hands are played, with dealer playing last, gives the casino an advantage of something over 5 percent.

Baldwin realized that working out all the permutations and combinations of the various hands was going to be a serious matter. He enlisted his fellow soldiers, Sergeant Wilbert Cantey; Herbert Maisel, who also had a Master's degree in mathematics and later taught at Georgetown University; and finally James McDermott who, like the rest, had a Master's degree in mathematics. McDermott and Maisel were also draftees. Cantey had been in the reserve, which is why he was a sergeant. The group got permission to use the base's calculators after hours. When I recently talked to Maisel he told me that Baldwin was definitely their leader. The "four horsemen," as they are known to blackjack aficionados, worked in their spare time for the next year and a half perfecting their strategy, which resulted in the article mentioned before. Maisel told me that they were never aware of anything that went on at Los Alamos, and he was surprised when I told him about it. Maisel never used the scheme in a casino, but Baldwin did. He told me that he spent a week or so in the summer of 1954 in Las Vegas seeing what the actual casino rules were. He visited all sixteen that then existed. As I will note later, Baldwin found that the "horsemen" had an important point wrong and had to redo some of their calculations.

The scene now shifts to California. Robert Sorgenfrey, who was a mathematics professor at UCLA and an avid bridge player, had come across the Baldwin et al. article. He knew a young mathematics instructor named Edward Thorp, a former physics major who had switched to mathematics and taken his Ph.D. at UCLA, Thorp had decided to take a Christmas vacation with his wife in Las Vegas. The Thorps were not gamblers but had been to Las Vegas before, since the prices of hotels and meals were reasonable. Sorgenfrey called Thorp's attention to the new blackjack strategy, and Thorp decided to try it out. He purchased 10 silver dollars and after some considerable efforts at the table, lost most of it and quit. When he got back to UCLA he decided that the strategy needed improvement. During the summer he moved to MIT, where he became an instructor, and it was there that he devised the improved strategy.

Thorp's advance changed the way casino blackjack is played. He realized that if you altered your bet as a function of what cards had been played in previous hands, you could alter the odds in your favor. This, of course, required card counting. First

I will explain how this differs from the basic strategy of Baldwin et al. To explain their strategy, which does not involve this sort of card counting, I will consider a situation of one player and a dealer using one deck. I should note that the last time I played blackjack was in an Indian casino along the superhighway that now leads from Santa Fe to Los Alamos. It was fairly early in the morning, and I was the only player. The difference was that the dealer used several decks that are dealt out of what is known as a "shoe." I did not last long.

Ignoring the matter of the burned card, when the deal is done you have two cards and the dealer has two, with one of the cards face up. The knowledge of these three cards is all the information you have. You must now make one of four choices. Two of the choices I have already mentioned – staying put or taking a hit. But there are two others.[3] If you receive two like cards you can "split" them. The cards are turned face up, and the dealer puts a new card face down on each member of the pair. You bet an amount equal to our original bet on the new hand, and the two hands are played separately. The basic strategy tells you when to split as a function of what the dealer is showing. Some of the rules are obvious and some aren't. Clearly, if you have two aces you always split them. But if you have two tens you never split them. You are already beginning with a sum of 20 and a decent chance of beating the dealer. Much more subtle and counterintuitive is that you always split two 8's, whatever the dealer is showing. This is the sort of thing that only becomes evident when all the possibilities are worked out and shows why you need some heavy computation.

The other choice you can make is "doubling down." If you think that the two cards you have look promising, for example they may add up to 11, you may double your bet and take one and only one additional card from the dealer. Until Baldwin's trip the "horsemen" thought that doubling down was only allowed if you were dealt a total of 11. The basic strategy tells you when, in general, to double down. You always do it when your cards add up to 11 and you never do it, or almost never do it, if your cards add up to what is called a "hard 8" – an 8 that is realized with no ace present. Again you see that working out all the possibilities as a function of what the dealer's face-up card shows is not a trivial matter.

Once these decisions are made you are ready to play your hand, or hands. The basic strategy tells you what to do as a function of what you are holding and the dealer is showing. For example, if the dealer is showing a 7 and you have a hard 17 or more, you stand, and so on. If you do all this, and live by the system, you can beat the house by a fraction of a percent. A lot of work for rather little reward, at least financially. This is what Thorp thought.

However, he realized that in the course of an actual game there was information that was revealed that the basic strategy ignored. If you took advantage of this information and changed your bet accordingly from hand to hand you could actually beat the house. In the course of the game what is revealed are the cards that

---

[3] There is in some casinos a third which is called "insurance." This is a hedge against the dealer getting a blackjack. In the interests of simplicity I will not consider it here.

have been played as well as the number of cards that remain to be played. Again let us assume that one deck is being used. If you see that the four aces have been played, you know that in the next hand you cannot draw an ace, which lowers your chances, so you might want to lower your bet. (The lack of aces lowers your chances by something like 3 percent.) If all the aces are still available you might want to raise your bet, and so on. We are now entering the realm of card counting.

One might at first think that card counting would require an eidetic memory. That would help, but there are simplifications that enable mere mortals to play the system. In a general way low cards are friends of the dealer, who must take a hit on any total of 16 or less. The friendliest of all the cards are the 5's. There is no way that the dealer hitting on 16 or less can go over 21 by drawing a five Thus the simplest card counting system of all is to count the 5's that have been played along with the number of cards that remain. The latter requires more mental horsepower than the former. More sophisticated is to assign a number to each card that has been made visible. Cards from 1 to 6 are assigned plus 1, cards 7, 8, and 9 are counted 0, and the rest are assigned minus 1. After each round the counter adds up the score and, keeping track of the remaining cards, bets accordingly.

Viewers of the film *21,* which was loosely based on the book *Bringing Down the House* by Ben Mezrich,[4] will recall the scene in which the protagonist, an MIT student called in the film Ben Campbell, played by Jim Surgess – the real Ben Campbell was an MIT student named Jeffery Ma – is being indoctrinated by an unsavory and, in this instance, an entirely fictional, MIT professor played by Kevin Spacey. (MIT refused to allow the film to be shot on its campus, so Boston University was used.) The indoctrination consists of Kevin Spacey shouting amid a background of general chaos numbers of the count, which Ben Campbell is supposed to keep straight. It went by so fast in the film that I could not make heads or tails of what was going on. Back to Edward Thorp.

By June of 1960 he had the strategy worked out. He presented a paper at a meeting of the American Mathematical Society in Washington, DC, which he called "Fortune's Formula: A Winning Strategy in Blackjack." This was a combination of the basic strategy and counting the 5's. Before he left for Washington he received a phone call from a reporter at the *Boston Globe* who somehow had seen the abstract and wanted to do a story. The *Globe* even sent out a photographer. One must understand that blackjack is the most popular gambling game in casinos and that there are a lot of casinos, more all the time. (The notion that there *is* a winning strategy in blackjack is like announcing that there is gold in California. Most people don't understand that these strategies are statistical and that winnings of a few percent are possible if you are prepared to play for a long time, during which you have a good chance of losing all your money.)

Thorp reports that after his talk in Washington he was asked to give a press conference, which was followed by radio and television interviews and then hundreds

---

[4] Free Press, NY, 2003.

of letters and phone calls containing all sorts of offers. He finally chose offers from two professional gamblers, which he calls in his book, *Beat the Dealer*,[5] "Mr. X" and "Mr. Y." The three of them headed to Las Vegas with X and Y supplying a $10,000 grub stake. In his book *Fortune's Formula*[6] William Poundstone reveals that 'X' and 'Y' are respectively Manny" Kimmel and Eddie Hand. Kimmel was a shady bookie who made a fortune in parking lots and funeral homes. Hand was a gambler, then well known in Las Vegas casinos. It is very unlikely that either one had any understanding of the mathematics of what Thorp was doing.

By this time Thorp had improved his system. By June of 1960 he had had the basic information from which to construct all of the principal counting strategies that later became prominent. In particular, he had the 10's strategy worked out. It was picked it first because of the large number of 10's in the deck. He didn't present it in his Proc. N.A.S. paper dated January 1961, because there wasn't room; but he had alluded to it. At the January 1961 AMS meeting he only wanted to show proof of a winning system, and the 5's strategy was sufficient as well as being very simple and very startling. But now, instead of counting 5's, he was counting 10's. The presence of 5's substantially reduces the dealer's chances, while the presence of 10's somewhat enhances the player's. But there are sixteen 10's and only four 5's, so the cumulative effect of counting 10's is better.

Thorp gives an amusing account of his stay in Las Vegas with X and Y. It fell into a pattern. When he started to win the casinos simply kicked him out or changed the rules so as to diminish his chances, for example, by introducing new decks in mid-play. This slows the game down, so the casino is not anxious to do this very often, but it is better than being cleaned out. The casinos did not give reasons for throwing him out as a rule. They felt that something was going on that they did want to be a victim of. Unlike in the movie, where there is a great deal of physical violence as the card counters get beaten up, this was done in Thorp's case with the greatest of good manners. He and X and Y were sometimes offered free dinners and unlimited drinks but told to go away. The film is also unrealistic in that huge sums of money seem to change hands at lightning speed. In Thorp's case the maximum that was ever at stake in a single game was about $1,000. At the $500 limit tables he sometimes had two or three multiples of $500 down, either from pair splitting, doubling down, multiple hands, or a combination of these. Undoubtedly, at least some of the times one or several of the four $500 bets was at risk. Nonetheless, the three of them came away with a profit of some $11,000. This was peanuts compared to what Thorp must have made on his book, which was a national best seller. A second edition appeared in 1966. Between the editions a computer scientist at IBM named Julian Braun ran the strategy on the latest IBM main frame computer. He used Thorp's annotated program and suggestions to write a revision that took the calculations out to greater accuracy, possible because of the increase in computer power in the intervening years. It was only

[5] Vintage Books, New York 2nd edition 1966. The first edition was published in 1962. I would like to thank Ed Thorp for many helpful comments
[6] Hill and Wang, New York, 2005

20 years after Thorp's initial work that Peter Griffin, with the more advanced computers of the day, was able to an exact calculation for the first time. I do not know if it produced better results than the MANIAC at Los Alamos, but Thorp used Braun's results in his 1966 edition. Thorp's book has now sold more than 600,000 copies. Incidentally, early in the movie, a copy of the book is shown on a table in passing.

As I have mentioned, the film *21* is loosely based on Ben Mezrich's book *Bringing Down the House*, whose title has now morphed into *21: Bringing Down the House*. The book, which purports to be non fiction, is also loosely based on reality. Before I get into that let me describe the apparent facts. MIT has something called an Independent Activities Period, which encourages its students to do things outside the curriculum. I think of this as a somewhat more adult version of the programs designed to keep kids off the street by offering them after-school activities. When I was a student at Harvard I took a few courses at MIT, and nothing would surprise me about MIT undergraduates, or so I thought. In January of 1979, such a course was offered called "How to Gamble if You Must." This must have seemed like a wonderful idea to someone. If you are going to practice a vice, at least do it intelligently.

Several MIT students attended this course and learned about card counting in blackjack. Flush with this knowledge some of them headed to Atlantic City and got their clocks cleaned. Most of the group decided that this was enough, but a couple persisted and even taught a second version of the course in January of 1980. About this time one of them, J. P. Massur, had an accidental meeting in a Chinese restaurant with a recent Harvard business school graduate named Bill Kaplan, who had in fact been running his own blackjack team in Las Vegas. In Mezrich's book – one is tempted to say novel – Kaplan has become Micky Rosa, an MIT graduate and former instructor. In the film he has become a full professor of mathematics, in effect, Kevin Spacey. (The real Bill Kaplan looks about as much like Kevin Spacey as the film resembles real events, to say nothing of the book.)

Kaplan agreed to come with the MIT group on a second foray to Atlantic City. He did not do this for intellectual reasons. Kaplan was making a business of sponsoring blackjack teams and had just parted ways with his previous team. What he observed in Atlantic City was another disaster. Each member of the team seemed to have an individual counting strategy and spent a great deal of time arguing about which was best. Kaplan said that he would back the team provided that they would change their modus operandi. For example, there would be one counting system and that would be the one that Kaplan told them to use. The new team, with a capitalization of $89,000, started in August of 1980, and by two months later the team members were earning about $80 an hour of play while Kaplan and his other investors were raking it in by the carload.

Oddly, in reading what I have been able to find, I do not see that Jeff Ma, Kevin Lewis in the book and Ben Campbell in the movie, played an especially important role. He was certainly never a medical student, as the film portrays. He seems to have taken the money he earned and bought a town house in the South End of Boston and invested in a bar. The MIT team grew to some eighty players before it broke up in 1993. By this time most of them had been banned from most casinos, and the casinos were using multiple decks hidden in a "shoe." Some of the team went into real estate, thinking there was more money to be made there. Maybe they have now returned to blackjack.

As for Thorp, he became a professor of mathematics, ending his academic career in 1983 at the University of California at Irvine. By this time he had employed the skills he had acquired in the study of blackjack to the stock market. He became associated with various hedge funds, which managed hundreds of billions of dollars. He always beat the market, and in some years he even beat Warren Buffet. Perhaps the best known of these funds was the one he founded with a partner, James Regan, called the Princeton-Newport Partners. Regan was in Princeton and Thorp in Newport Beach, California. In its heyday the fund managed money for institutions such as Harvard. Its heyday came to an abrupt end in 1988 when the fund was investigated by Rudy Giuliani. Giuliani threatened to use the then-new RICO law to indict the company. It turned out that the Princeton wing had been engaged in "stock parking." This is the practice of holding stock for someone else so that tax advantages can be gained. In this case the someone else turned out to be people such as Ivan Boesky and Michael Millken, both of whom ended up in jail. The company was never indicted, only threatened with indictment if Regan wouldn't "rat out" Milken and Freeman. He didn't, and there was a trial of five people from the Princeton office. Thorp had had nothing to do with all this, but the fund was dissolved. Since then he has run his own funds.

I have never understood the casino paranoia about "counters." It contrasts vividly with what happens with the slot machines. If someone hits the jackpot the casino practically declares a national holiday. They want people to know that there are winners so that the flotsam and jetsam are sucked in. Somehow they feel differently about blackjack. The huge majority of blackjack players do not understand the system and do not realize how fast the casino game moves and how much distraction there is. They also do not understand the statistical nature of the system. As an illustration I will end this with an absolutely guaranteed way of making money in a casino.[7] Go to a roulette wheel and put down $2 on red. If it comes up red, you have won $2. If it comes up black, put $4 on red. If it comes up red you have won $2. If it comes up black, put $8 down, and so on. If the wheel is "rational," it eventually has to come up red. But keep in mind that $2 raised to the 20th power is $1,048,576. To paraphrase what Lord Keynes might have said: "Roulette wheels can remain irrational longer than you can remain solvent."

---

[7] This is an example, indeed the paradigmatical example, of what in probability theory is known as a "martingale strategy." In an actual casino there are one or more green places on the wheel which do not count as red or black. Moreover there are house limits to betting that make it easier for the punter to get wiped out by fluctuations. Variants of the martingale strategy are used by hedge fund managers.

Index

A

Abdus Salam International Centre for
    Theoretical Physics, 48
Aberdeen Proving Ground, 162
Académie Française, 149
Advanced Research Projects Agency
    (ARPA), 40
Afghanistan, 48, 49, 54
African Black Congress, 44
Alchemy, 67
Alexander the Great, 91
All is true (play), 66
"All That Glitters" (Bernstein), 87–105
Almelo, Treaty of, 52
Althorp Park, 75
Amaranth Advisors, 22
American Heritage dictionary, 108
American Journal of Archaeology, 89
American Mathematical Society, 165
American Revolution, 82–83
American Scholar, The, 155
Analysis situs (algebraic topology), 148
"An Application of the Poisson Distribution"
    (Clarke), 124, 125
Anderson, Philip, 7
Angular momentum, 127–128
Anne, Princess of Denmark, 59
Anti-Semitism. *See also* Jews, Polish; Nazi
    party
  of H. P. Lovecraft, 135
  of Hans Frank, 28
  in Poland, 116
Arbitrage, effect on stock option prices, 11
Architectural Association School, London, 89
Aristotle, 108
Arthur D. Little Company, 15, 16
"arXiv," 150–151
Asiatic Researches, 91
Asiatic Society, 85, 91

Aspect, Alain, 121, 127–128, 131, 132, 137
Astrology, 68
Astronomy, planetary, 67–68
Atomic bomb, 36–37
  "boosted," 50–51
  "gun assembly" device design for, 43, 44
  Hiroshima and Nagasaki bombings,
    38, 43, 55
  Manhattan Project, 52
  as space vehicle propellants (Project
    Orion), 35–43
Atomic bomb tests, 161
Atomic Energy Board (South Africa), 43–45
Atomic Energy Commission (Pakistan),
    48, 51, 53
Atomic Energy Commission (United States), 37
Atomic physics, 32–33
AT&T Bell Labs
AT&T Bell Labs. *See* Bell Labs
Auschwitz concentration camp, 115, 119
Au Vieux Campeur, 138

B

Bachelier, Louis, 12, 15, 159, 160
Bacon, Edmund, 66
Bacon, Francis, 58, 60–61, 66, 67
Baldi, Ottavio, Henry Wotton as, 58–59, 65
Baldwin, Roger, 161, 162, 163–164
Balliol College, Oxford University, 64
Ballistic missiles, V-2 rockets as, 124
Bamiyan, Afghanistan, giant Buddha
    statutes of, 48
Bangladesh, 53
Bankers Trust, 17
Bank of England, 82
Bardeen, John, 7
Barthes, Roland, 135–136
Bass, Robert, 22

B.B.C. (British Broadcasting Company), 101
BBC China (ship), 54
Beams, Jesse W., 52
Bear Sterns, 21
"Beating the System" (Bernstein), 159–168
Beat the Dealer (Thorp), 166, 167
Beckett, Samuel, 135–136
Beg, Mirza Aslam, 47
Bell, Beverly, 17
Bell, John, 127, 128–132, 132, 155
    "Bell's inequality," 130–131, 139–140
    "Bertlmann's Socks and the Nature of
        Reality," 127
    FAPP ("For All Practical Purposes")
        concept of, 140, 143
Bellarmine, Roberto, 68
"Belle" (chess-playing machine), 7
Bell Journal of Economics and Management
        Science, 15
Bell Labs, 4–5, 7, 17, 19
    Business Analysis Systems Center, 6–7
"Bell's inequality," 130–131, 139–140
Bengali language, 84, 93
Bennett, E. L., Jr., 97
Berenson, Bernard, 63
Beria, Lavrentiy, 53
Berman, Jakub, 120
Bernstein, Jeremy
    "A Cosmic Flow," 155
    "All That Glitters," 87–105
    "A Nuclear Supermarket," 47–55
    "Beating the System," 159–168
    Continuous Opacity and Equations
        of State of Light Elements
        at Low Density, 40
    Dawning of the Raj, 90
    "In a Word, 'Lions' ", 107–110
    "Options," 3–7
    "Orion," 35–42
    "Ottavio Baldi:The Life and Times of Sir
        Henry Wotton," 57–69
    Publication of the Astronomical Society
        of the Pacific, Vol. 115 (2003)
        1383–1387, 39
    "Rocket Science," 121–132
    "Tales from South Africa," 43–45
    "The Pianist, Fiction and Non-fiction,"
        113–120
    "The Rise and Fall of the Quants," 15–22
    "The Spencers of Althorp and Sir William
        Jones: A Love Story," 73–110
    Three Degrees Above Zero, 6–7
    "Topology," 147–154
    "What the #$*!?", 155–157
Bertlmann, Reinhold, 127, 129–130

"Bertlmann's Socks and the Nature of
    Reality" (Bell), 127
Bethe, Hans, 150
Bhagavad-Gita, 84
"Bhangmeter," 44
Bhutto, Zulifikar Ali, 53
Biermann, Wolf, 120
Big Bang, 6–7, 143
Bingham, George Charles (Earl of Lucan), 78
Bingham, Lavinia, 78
Black, Fischer, 4, 15, 16–17
    "Interest Rates as Options," 17
Blackjack (card game), 161–168
Black-Scholes equation, 4, 7, 9–14, 19
    first derivation of, 15
    as heat diffusion equation, 13
    as "random walk" model, 12–13
    "toy" model application of, 10–11
Blackstone, William, 79
BLAND Corporation, 35
Blegen, Carl W., 94–95, 102
Boesky, Ivan, 168
Bogucki, Andrzej, 117
Bogucki, Janina, 117
Bohm, David, 129, 139–140
Bohr, Niels, 29, 31, 130, 131, 149
Bond option pricing, 4, 16
    BDT model of, 17
Boness, A. James, 15
Bonfante, Giuliano, 107
Bonfante, Larissa, 107
Born, Max, 29, 140
Boston Globe, 165
Boston University, 131, 165
Boswell, James, 80
Botha, P. W., 44
Botticelli, Sandro, 108
Bouillon de Culture (television program), 121
Boyle, Charles (Earl of Cork), 67
Boyle, Robert, 67
Brabers, M. J., 51
Brattain, Walter, 7
Braun, Julian, 166–167
Bringing Down the House (Mezrich), 165, 167
British Atomic Energy Research
        Establishment, 129
British East India Company, 82, 83
British School of Archaeology, Athens, 89
Brokerage firms. See also names of specific
        brokerage firms
    construction of synthetic options by, 11
    financial engineering departments of, 4, 7
Brooklyn College, 94
Brown, Robert, 12
Brownian motion, 12–13, 15, 19, 159

Buchenwald concentration camp, 123
Buddhism, 155
Buffet, Warren, 21, 168
Bühler, Josef, 28
Burke, Edmund, 80, 81, 84

C

Caïssa, or The Game of Chess (Jones), 77
Calculators, 39, 163
Calculus
    derivatives in, 4
    invention of, 75
    stochastic, 12
Cambridge University, 48, 101, 145
Camera obscura, 61
Campus (television program), 121
Canada, involvement in Pakistan's nuclear
        program, 50
"Can Quantum Mechanical Description of
        Physical Reality be Considered
        Complete" (Einstein et al.), 131
Cantey, William, 161, 163
Capra, Fritjof, 155
Card-playing, 160–168
Catholic University of Leuven, Belgium, 51
Cecil, Robert (Earl of Salisbury), 67
Centrifuge technology, 44, 51–54
CERN, 127, 155
    particle accelerator of, 129
Chadwick, John, 101–103
Chaos theory, 148
Charles I, king of England, 59
Chen-Ning Yang, 5
Chess, 77, 91, 161
Chess-playing machines, 7, 171
Children of a Lesser God (film), 155–156
Chimaera of Arrezzo, 108
China, involvement in Pakistan's nuclear
        program, 50, 53
Christ, Norman, 6
Christian VII, King of Denmark, 78
Clarke, R. D., 124–126
Clauser, John, 131
Clay, T., 154
"Coarse graining," 143
Coblitz, Wilhelm, 30, 31, 33
Code breakers, 95–96, 98
Code of Gentoo Law, 84
Cold War, 35, 160
Collins, John, 75
Columbia University, 4–6, 22, 162
Compleat Angler (Walton), 64
Computers, 161, 166–167
Comus (Milton), 66–67

Concentration camps, 27, 119
    Auschwitz, 115, 119
    Buchenwald, 123
    Sachsenhausen-Oranienburg, 27
    Treblinka, 115
Condon, Joe, 7
*Continuous Opacity and Equations of
        State of Light Elements at Low
        Density* (Bernstein and Dyson), 40
Convergent infinite series, 150
Copenhagen, Denmark, 29, 59
Copernicus, Nicolaus, 68
Corera, Gordon, 52, 53, 55
Cornell University, 124, 151
Cornfeld, Bernie, 18, 19
Cornwallis, Charles, 83
Corzine, Jon, 19
"Cosmic Flow, A" (Bernstein), 155
Cosmic rays, 32–33
Cosmology, 68
"C" programing language, 7
Cracow, Poland, 25–26, 27
Crete
    Linear A and B language tablets from,
        87–88, 89, 93–94, 100
    Mycenaean colonization of, 94
Crocodile (ship), 84
Culture, scientific, 122–123
Curie, Marie, 147
Curved spaces, geometry of, 152

D

Dalai Lama, 155
Dawning of the Raj (Bernstein), 90
Death of a City (Szpilman), 113
De Bessy, Bernard Frénicle, 146
Degussa Company, 52
De Hoffman, Frederic ("Freddy"), 35–36, 37
De Klerk, F. W., 44
Denmark, Germany's invasion of, 116
Derivatives, stock and bond options as, 4
    definition of, 9–10
    "in the money," 10
    "out of the money," 10
    "strike prices" of, 9–10
Derivative trading, 19
Derman, Emanuel, 4–7, 22
    at Bell Labs, 4–5
    at Columbia University, 4–5, 7
    at Goldman Sachs, 4, 7, 15, 16, 17
    "In the Penal Colony," 6
    *My Life as a Quant: Reflections on
        Physics and Finance,* 4–5
    as University of Cape Town graduate, 5

Desplechin, Arnaud (fictional character),
    121, 127
Deuterons, 50
Devereaux, Robert (Earl of Sussex), 58
Devis, A. W., 80
Dirac, P. A. M., 140–141
Djerzinski, Michel (fictional character),
    121, 126–127, 132, 137
Dr. Strangelove (film), 35
Donne, John, 57–58, 59
    biography of, 62
    Ignatius His Conclave, 60
    meeting with Johannes Kepler, 60
Dormobile, 48
Dornberger, Walter, 123
Dravidian languages, 93
Dreyfus, Alfred, 149
"Drunkard's walk," 12–13, 159
Dubai, 51
Durand, Guillaume, 121, 132, 136–137
Dynely, John, 65
Dyson, Esther, 140
Dyson, Freeman, 11, 95, 140, 150
    at General Atomics, 35
    Project Orion and, 36, 37, 38, 39–40
    TRIGA reactor and, 36
Dyson, George, 140
    Project Orion, 36, 38, 40, 41, 42

E
East India Company, 82, 83
Ecclesiastus (Scioppius), 59
École des Mines, 148
École Louis-Lumière, 135
École Normale Supérieure, 134
École Polytechnique, 148, 149
Economics
    mathematical, 15
    role of physics in, 158
"Efficient market hypothesis," 159
Einstein, Albert, 28, 130, 131, 147, 150, 159
    Brownian motion and, 12, 13
    "Can Quantum Mechanical
        Description of Physical
        Reality be Considered
        Complete," 131
    on nature of reality, 141
    quantum theory and, 139–140, 141
    theory of general relativity and gravitation,
        22, 152–153
    theory of relativity, 147, 152
Electrons, two-slit passage of, 156–157
Elementary particle physics, 142, 160–161

Elementary Particles, The (Houellebecq),
    121, 126–127, 132, 134, 137,
    139, 140
Elements of Architecture, The (Wotton), 68
Elizabethan literature, 'great age' of, 64
Elizabeth (daughter of James VI), 59, 60
Elizabeth I, Queen of England, 58, 66
Élohimites, 139, 144
Enceladus (moon of Saturn), 38
Enigma code, 95–96
Enron, 18
Epp Freikorps, 26
Equator, 152
Eton College, 57, 58, 59, 62, 64, 67, 68
Etruria, 108
Etruscan language, 89–90, 99–100, 107–110
Etruscan Language: An Introduction
        (Bonfante and Bonfante), 107
Etruscans, 107
Euclid, 79
Evans, Arthur, 88, 89, 97, 98, 99, 101
Everett, Cornelius, 36–37
"Evidence for Greek Dialect in the
        Mycenaean Archives" (Ventris
        and Chadwick), 102
Express Livre, 136
"*Extension du Domaine de la Lutte*"
        (Houellebecq), 135–136

F
FAPP ("For All Practical Purposes") concept,
    140, 143
FBI (Federal Bureau of Investigation), 37
Federal Reserve, 18, 21
Feinberg, Gerald, 6
Feklisov, Alexander, 55
Fellig, Arthur, 35
Ferdinando I, Grand Duke of Tuscany, 58
Fermi, Enrico, 49
Feynman, Richard, 11, 141
Fields Medal, 153–154
Financial Analysts Journal, 17
Financial engineering, 22
Financial engineering departments,
        quants in, 4, 7
Fission reactions, 49–50
Fission weapons, 44
Flammarion, 136, 1399
Florence, Italy, 67
    Grand Duke of, 65
"For All Practical Purposes" (FAPP) concept,
    140, 143
Ford Foundation, 3, 47, 48

"Fortune's Formula: A Winning Strategy
    in Blackjack" (Thorp), 165
Fortune's Formula (Poundstone), 166
Fourth dimension, 153
Fox, Charles James, 80, 82
France, Germany's invasion of, 116
Frank, Hans, 26, 27, 28, 30, 31, 32, 33, 116
Frank, Niklas, 27
Franklin, Benjamin, 81
Frederick Lord North, 82, 83
Frederic V, King of Bohemia, 59
Freedman, Michael, 151
Friedman, David, 5
Friedman, Milton, 5
Fuchs, Gerhard, 55
Fuchs, Klaus, 53, 55
Full Tilt (Murphy), 47–48

G

Gadafi, Muammar, 54
Galbraith, John Kenneth, 19
Galileo Galilei, 67–68
    *Siderius Nuncius (The Starry
        Messenger),* 67
Gambler's prayer, 161–162
Gambling, theory of, 160
Garrick, David, 80
Gell-Mann, Murray, 142–143
General Atomic
    Project Orion, 35–45
    TRIGA reactor, 36
General Atomics, 35
General Dynamics, 35
"Genus one," 152
Georgetown University, 163
Gerlach, Walther, 128
Germany, invasion and occupation
        of Poland
    Warsaw Ghetto, 113–120
    Werner Heisenberg and, 25–34
Gestapo, 27
Gibbon, Edward, 73, 80
Gimpel, Bronislav, 120
Gingerich, Owen, 61
Ginsparg, Paul, 151
Giuliani, Rudy, 168
Glashow, Sheldon, 48
Global Crossing, 18
Globe Theater, 66
Goldman, Henry, 16
Goldman Sachs, 4, 7, 15, 16, 17, 19
Goldsmith, Oliver, 80
Gosse, Edmund, 62–63

Gowing, Margaret, 29
"Graining," 142–143
*Grammar of the Persian Language,
    A* (Jones), 80
Gravitation, theory of, 22, 152–153
Gravitational Rainbow (Pynchon),
        121, 122–124, 126, 132
Greece, ancient, Proto-Indo-European people
    in, 93–94
Greek language
    relationship to Linear B, 100–102,
        103–104
    relationship to Proto-Indo-European
        language, 92–94
    relationship to Sanskrit language,
        90–92, 103
Greenspan, Alan, 20–21
Griffin, Peter, 166–167

H

Hahn, Otto, 49
Hamilton, Richard, 152, 153
Hand, Eddie, 166
Hapgood (play), 156
Hardy, G. H., 145–146
Harmony of the World (Kepler), 60, 61
Harrow School, 74, 75, 77, 78–79
Hartle, James, 142–143
Harvard Business School, 22
Harvard University, 3, 4, 16, 63, 149,
        160, 168
Hastings, Warren, 82, 84, 85, 90, 91, 95
Hawkins, Stephen, 142
Heat diffusion equation, 13
"Heavy hydrogen," 49–50
Hedge funds, 18, 19–22, 168
Heisenberg, Elisabeth, *Inner Exile,* 25, 29–30
Heisenberg, Erwin, 26
Heisenberg, Werner
    at Copenhagen, 31
    German nationalism of, 29
    in Nazi-occupied Poland, 25–34
    as Nobel Prize winner, 29
    as student in Maximillians ("Max")
        Gymnasium, 26
    "Unity of the Scientific Worldview," 31
    as "White Jew," 30
"Heisenberg in Poland" (Bernstein), 25–34
"Helikon tubes," 43–44
Henry VIII, king of England, 66
Heydrich, Reinhard, 28
Hickey, William, 86
Himmler, Heinrich, 30, 31

Hindi language, 93
Hindu law, 84, 86
Hindustani language, 84
Hiroshima, atomic bombing of, 38, 43
Histoire de Mader Shah (Jones), 78
Hitler, Adolf, 26, 27, 28, 116, 117, 119, 135
Holt, Richard, 131
Homer, 94, 100, 101
Horne, Michael, 131
Hosenfeld, Wilm, 119–120
Houellebecq, Henriette, 134
Houellebecq, Michel, 121–123, 133–146
    birth of, 133
    education of, 134–135
    "Extension du Domaine de la Lutte",
        135–136
    grandmother of, 134, 135
    "H.P. Lovecraft: Against the World,
        Against Life," 135
    La Possibilitié D'une Île, 133, 139,
        143–144, 146
    Les Particules élémentaires, 121, 126–127,
        132, 134, 137, 139, 140
    marriage of, 135
    parents of, 133–134
    Plateform, 138–139
"H.P. Lovecraft: Against the World, Against
    Life" (Houellebecq), 135
HTML (hypertext markup language), 133
Hydrogen bomb, 36, 50

I
IBM (International Business Machines),
    6, 166
Ignatius His Conclave (Donne), 60
IGUS ("information gathering and utilizing
    system"), 143–144
Imperial College, London, 48
Impey, Elijah, 82
"In a Word, 'Lions'" (Bernstein), 107–110
India
    British governance of, 82
    languages and dialects of, 84, 93
Indian Mathematical Society, 145
Indo-Aryan languages, 93
Indo-European languages, 85–86
Inner Exile (Heisenberg), 25, 29–30
"Institute A," 52
Institute for Advanced Study, Princeton,
    35, 40, 95, 160
Institute for the Study of Twins, 132
Institute nationale agronomique (INA),
    134–135

Institut für Deutsche Ostarbeit (Institute
    for German Work in the East),
    30, 32, 33
Institutional Investor, 18
"Interest Rates as Options" (Black), 17
International Atomic Energy Agency (IAEA),
    44, 45, 53–54
International Overseas Services, 18
Internet, 150–151
"In the Penal Colony" (Derman), 6
"Introducing the Minoan Language"
    (Ventris), 89
Iran, 48, 51, 53–54
Iraq, 52
Irving, David, 33
Islam, 138, 139
Islamabad, Pakistan, 48
    university of, 3, 47, 48–49
Israeli, 44

J
Jaccoux, Claude, 48
Jaccoux, Michele, 48
James, William, 63
James I, King of England, 66, 68
James VI, King of Scotland, 58–59, 65–66
Jews, Polish
    Nazi extermination policy toward,
        27, 29–30, 113, 114, 119–120
    in the Warsaw Ghetto, 113–120
Johnson, Samuel, 73, 80, 81
Jones, Mary, 75
Jones, William (father), 74–75
    *A New Compendium of the Whole
        Art of Navigation,* 74
    Synopsis palmariorum matheseos, 77
Jones, William (son), 73–86
    as Asiatic Society founder, 85, 91
    biographies of, 86
    Caïssa, or The Game of Chess, 77
    as candidate for Parliament, 81–82
    death of, 86
    A Grammar of the Persian Language, 80
    in India, 73–74, 76, 77, 79–80, 82–83,
        84–85
    knighthood of, 84
    language proficiency of, 74, 77
    law practice of, 80, 81
    law studies of, 80
    letters to George John Spencer, 73–74,
        76–77, 79, 80–81, 83, 84–85
    L'Histoire de Mader Shah, 78
    as Literary Club member, 73, 80

marriage of, 84, 86
The Muse Recalled, 78
as Orientalist scholar, 80, 81, 84, 85
    Proto-Indo-European language studies,
        85–86
    Sanskrit language studies, 84, 85, 86,
        91–93
    parents of, 74–75
    as tutor to George John and Georgiana
        Spencer, 74, 75–79
Journal of Hellenic Studies, 102
Journal of Political Economy, 15
*Journal of the American Statistical
    Association,* 161
Journal of the Institute of Actuaries, 124
Judicature Act, 82
Jupiter, moons of, 67–68
JWM Partners, 22

K
Kahn, Herman, 35
Kaiser-Wilhelm-Institut für Physik, 32
Kalidasa, 90
Kaplan, Bill, 167
Kepler, Johannes, 67
    Harmony of the World, 60, 61
    meeting with John Donne, 60
    relationship with Henry Wotton, 60–61
Keynes, John Maynard, 15, 20, 160, 168
Khan, Abdul Qadeer (A.Q.), 45, 51–52, 53,
        54–55
Khan, Ayub, 47
Khan, Yahya, 47, 53
Khyber Pass, 48
Kimmel, Manny, 166
King, Stephen, 135
Knossos, Crete, 88, 89, 93–94, 98, 99,
        100, 103
Knowledge, "pure search for," 5–6
Kober, Alice, 94, 95, 97–98, 99, 100
Krakauer Zeitung, 31, 32
Kruizinga, Richard, 15
Krummacher, Annemarie, 119
Kubrick, Stanley, 35, 42

L
"Lagrangian in Quantum Mechanics,
    The" (Dirac), 140–141
Laika (dog), 37
Land Rovers, 48
Langevin, Paul, 147
La Nouvelle Revue de Paris, 133

Las Vegas, Nevada, 161, 163, 166, 167
Law of One Price, 11
Lawrence, D. H., 107
Lawrence Livermore National Laboratory, 160
Lehrer, Tom, 43, 44, 45
Lehr-Splawinski, Tadeusz, 27
Leibniz, Gottfried Wilhelm, 75
Leipzig University, 31
Lemaistre, Stephen Caesar, 82
Lemnian language, 108
Leningrad State University, 150
Leonides, 108
Lerch, Gotthard, 55
Leverage regulations, 20–21
L'Express Livre, 136
Libreria Francese, Florence, 121
Libya, 45, 51, 52, 54, 55
Liebling, A. J., 161
*Life and Letters of John Donne,
    The* (Gosse), 62
*Life and Letters of Sir Henry Wotton,
    The* (Smith), 61, 64, 65–68
Lilla, Mark, 137
Linear A, 93–94, 95, 104
Linear B, 89, 92, 93, 109
    deciphering of, 88–90, 94, 95–105
    discovery of, 87–88
    relationship to Greek language, 100–102,
        103–104
Linguistics, historical, 91
Linz, Austria, 60–61
Lion, etymology of, 107–108, 110
Lipperhey, Hans, 67
Lire, 138
Literary Club, 80
"Little Boy" (atomic bomb), 43
Livermore National Laboratory, 160
Lives, The (Walton), 58
Locke, John, 79
London, V-2 rocket attacks on, 122, 123,
        124–126
London Times, 103
Long Term Capital Management-LTCM,
        18, 19–22
Los Alamos National Laboratory, 36–37,
        43, 151, 160
    Jeremy Bernstein at, 160–163
    Q-clearance at, 160
    Richard Feynman at, 141
    Theory Division, 37
Lovecraft, H. P., 136
Low, Francis, 37–38, 161, 162
Lowenstein, Roger, 18, 20
Lucent, 7

M

Ma, Jeffery, 165, 167
Mackey, George, 4
Maisel, Herbert, 161, 163
Malaysia, 54
Man Behind the Rosenbergs,
        The (Feklisov), 55
Mandela, Nelson, 44
Manhattan Project, 52
MANIAC computer, 161, 167
Mark, Carson, 37–38, 161
Mars
    exploration of, 38, 42
    science fiction about, 122
Massachusetts Institute of Technology (MIT),
        16, 17, 161, 163, 165
    "How to Gamble if You Must" course, 167
    Independent Activities Period, 167
    Sloan School of Management, 15
Massur, J. P., 167
Matlin, Marlee, 155–156
Maximillians ("Max") Gymnasium, 26
Maxwell, James Clerk, 9
McDermott, James, 161, 163
"Mega-death," 35
Mercury, Nevada, 161
Meriwether, John, 7, 17, 18, 19, 21, 22
Merton, Robert (father), 7
Merton, Robert (son), 4, 7, 15, 18–19, 22
Mexico, Roger (fictional character), 126
Meyer, Johan, 45
Mezrich, Ben, 165, 167
Michel, Louis, 148
Milky Way *(Via Lactea),* 68
Miller, Henry, 135–136
Millken, Michael, 168
Milton, John, 64, 81
    Comus, 66–67
Minoan language. *See* Linear B
Moller, Christian, 29
Mont Blanc tunnel, 48
Monte Carlo Fallacy, 126
Moon(s)
    Galileo's observations of, 68
    of Jupiter, 67–68
    orbit of, 143
Mortgage crisis, 18
Mount Holyoke College, 148–149
Müller, Bruno, 27
Murphy, Dervla, 47–48
Muse Recalled, The (Jones), 78
Musharraf, Pervez, 54
Mycenae, 87
Mycenaean language, 87–88

Mycenaeans, 94
*My Life as a Quant: Reflections on Physics
        and Finance* (Derman), 4–5

N

Nadeau, Maurice, 135–136
Nadir Shah, 78
Nagasaki, atomic bombing of, 38, 55
NASA (National Aeronautics and Space
        Administration), 40, 42
National Archaeological Museum, Athens,
        87, 90, 93, 95, 102–103
National Polish Radio, 118, 120
Nazi party, 26, 28, 30, 119
    Adolf Hitler and, 26, 27, 28, 116, 117,
        119, 135
    Gestapo, 27
    "Night of the Long Knives," 26
    SS *(Schutzstaffel)* of, 27, 28, 114, 116, 123
    Storm Troopers, 26, 119
Nazri, "Tiger," 53
Nepal, 155
Nestor, Palace of, 94–95, 104–105
Netherlands, 51, 53
Neupfadfinders ("new pathfinders"), 26
Neutrons, in fission reactions, 49–50
*New Compendium of the Whole Art of
        Navigation, A* (Jones), 74
Newton, Isaac, 75
New Yorker, 48
New York Review of Books, 137
New York Times, 159
New York University, 107
"Night of the Long Knives," 26
Nix, Maria, 75
"Nobel Prize," for mathematicians, 154
Nobel Prize winners, 4, 5, 6
    Anderson, Philip, 7
    Bardeen, John, 7
    Brattain, Walter, 7
    Gell-Mann, Murray, 142
    Glashow, Sheldon, 48
    Heisenberg, Werner, 29, 33
    Merton, Robert, 7, 18
    Penzias, Arno, 7
    Salam, Abdus, 48
    Scholes, Myron, 7
    Shockley, William, 7
    Weinberg, Steven, 48
    Wilson, Robert, 7
Nocturne in C sharp (Chopin), 118
Norstad, John, 9, 10
North, Lord Frederick, 82, 83

North Korea, 51, 53
Northwestern University, 19
Nouvelle Revue de Paris, La, 133
Nouvelles Frontières, 138
Nuclear energy programs
    in Germany, 52
    in Pakistan, 49, 50
Nuclear reactors
    CANDU-CANadian Deuterium-Uranium, 50
    "heavy hydrogen," 49–50
    "heavy water," 50, 52
    KANUPP (Pakistan), 49, 50
    "light water," 50
    mechanisms of, 49
"Nuclear Supermarket, A" (Bernstein), 47–55
Nuclear test ban treaties, 41
Nuclear weapons. *See also* Atomic bomb;
        Hydrogen bomb
    boosted uranium implosion, 53
    directional, 38
    fission weapons, 44
    "gun assembly" device design of, 43
    "Plowshares" program and, 43
    Project Orion and, 36–42
    trafficking in, 45, 51–52 53–55
Nuclear weapons programs
    at Livermore National Laboratory, 160
    at Los Alamos National Laboratory,
        36–37, 43, 141, 151, 160 163
    in Pakistan, 48–52
    in South Africa, 43–45
Nuclear weapons technology, international
        trafficking in, 45, 51–52, 53–55
Numbers, theory of, 144–146
Number system, Mycenean, 87

O
*On a Method of Propulsion of Projectiles
        by Means of External Nuclear
        Explosions* (Ulam and Everett),
        36–37
Oppenheimer, Robert, 11, 150, 160
"Optimum Strategy in Blackjack"
        (Baldwin et al), 161
"Options" (Bernstein), 3–7
Orbital angular momentum, 127–128
"Orion" (Bernstein), 35–42
Oscar, King of Sweden, 148
O'Shea, Dona, 148–149, 150
Othello (Shakespeare), 66
"Ottavio Baldi:The Life and Times of Sir
        Henry Wotton" (Bernstein), 57–69
Oxford University, 63, 74, 75, 76–77, 78

P
Pais, Abraham, 141
Pakistan, 3–4
    Atomic Energy Commission, 48, 51, 53
    East (Bangladesh), 53
    first atomic bomb test, 53–54
    nuclear deterrence policy, 47, 53
    nuclear energy program, 49, 50
    nuclear weapons program, 48–52
    West, 53
Palm Pilot, 159–160
Parker, George, 74–75
Parker, Thomas (Earl of Macclesfield),
        74–75
Particle accelerators, 129
Particules Élémentaires, Les (Houellebecq),
        121, 122, 132, 134, 137, 139, 140
Pauli, Wolfgang, 29, 150, 156
Peierls, Rudolf, 19
Pension funds, 22
Penzias, Arno, 7
Perec, Georges, 135–136
Perelman, Grigori "Grisha," 149–152,
        153–154
Persian language, 84, 91, 93, 95
Pfadfinders ("pathfinders"), 26, 29, 30
Phoenicians, alphabet and language of,
        90, 96, 98, 100, 109
Photons, spin of, 127, 131
Physical Dynamic Research Laboratory
        (FDO), Almelo, Netherlands, 51
Physicists. *See also names of individual
        physicists*
    theoretical, ideal job of, 150
    on Wall Street (POWS), 4–7
Physics
    classical, 141
    relationship to nature, 130
Physics, 131
Pianist, The (film), 113–114, 116–117, 118
Pianist, The (Szpilman), 113–120
"Pianist, Fiction and Non-fiction, The"
        (Bernstein), 113–120
Pivot, Bernard, 121
Plagiarism, 151
Planck, Max, 142
Planck's constant (h), 142
Planets, orbits of, 67–68
Plateform (Houellebecq), 138–139
*Playing Blackjack to Win: A New Strategy
        for the Game of 21* (Baldwin et al),
        161
"Plowshares" program, 43
Plutonium, 38, 43, 50, 51

Plutonium reprocessing plants,
    in Pakistan, 51
Podolsky, Boris, 131
Poincaré, Henri, 147–149
    Science and Hypothesis, 149
Poincaré's conjecture, 151, 152
Poincaré's Prize (Szpiro), 148–149, 150
Pointsman, Edward W., 126
Poisson, Siméon-Denis, 124–125
    *Recherches Sur La Probabilité Des
        Jugements en Matière Criminelle
        et en Matiére Civile, Precedes
        Des Règles Générales du Calcul
        des Probabilities,* 124
Poisson distribution, 125
Poland
    anti-Semitism in, 116
    German invasion and occupation
        of, 27–28, 113, 116
    Warsaw Ghetto
        deportation of Jews from, 115–116
        escapes from, 116–118
        Jewish Council of, 116
        living conditions in, 114–115
        organization of, 114, 116
        underground army in, 117
        uprising in, 117
        Wladyslaw Szpilman's experiences
            in, 113–120
Poland, German invasion and occupation of
    Warsaw Ghetto, 113–120
    Werner Heisenberg and, 25–34
Polanski, Roman, 113, 116–117
Polarization, direction of, 131
Poles, Nazi prejudice toward, 116
Polish National Radio, 114
Possibilitié D'une Île, La (Houellebecq),
    133, 139, 143–144, 146
π, 75
Prentki, Jacques, 115
"Pricing of Options and Corporate Liabilities"
    (Black and Scholes), 15
Princeton-Newport Partners, 168
Princeton University, 141
Probabilities
    Poisson's formula for computation
        of, 124–125
    in quantum theory, 130
Probability theory, 12
Project Orion, 35–42
Project Orion (Dyson), 36, 38, 40, 41, 42
Proto-Indo-European (PIE) language, 85–86,
    92–94
Proto-Indo-European (PIE) people, 102

Protons, in fission reactions, 49–50
Proust, Marcel, 137
Prudential Assurance Company, Ltd., 124
Punjabi language, 93
Pylos, 94–95, 99, 100, 102, 104
Pynchon, Thomas, 124, 125, 138
    Gravitational Rainbow, 121, 122–124,
        126, 132

Q

Q clearance, 37
Quanta, of radiation, 143
Quants, 15–22
    definition of, 4
    recruitment into financial engineering
        departments, 7
    "The Rise and Fall of the Quants," 15–22
Quantum electrodynamics, 11, 13
Quantum mechanics, 32, 127, 128,
    129–130, 155
    "paths" in, 141–142
    unobserved phenomena in, 156
Quantum theory, 4, 122, 129, 130, 155
    Einstein and, 139–140, 141
Quark and the Jaguar, The (Gell-Mann),
    142, 143–144
Quarks, 142
Queen's College, 128–129

R

Rabi, I.I., 3, 5
Radcliffe College, 63
Radiation
    Cerenkov, 36
    quanta of, 143
Radioactive fallout, 41
Raetic language, 108
Ramanujan, Srinivasa, 144–146
RAND Corporation, 35, 39–40
"Random walk" model, of stock prices,
    12–13
Rao, Ramachandra, 145
Rawalpindi, Pakistan, 48
*Recherches Sur La Probabilité Des Jugements
    en Matière Criminelle et en Matiére
    Civile, Precedes Des Règles
    Générales du Calcul des
    Probabilities* (Poisson), 124
Regan, James, 168
Regulatory Act of 1773, 82
Relativity, theory of, 147, 152
Reliquiae Wottonianae (Walton), 62

Retirement funds, 22
Review of Economics and Statistics, 15
Reynolds, Joshua, 80
Ricci analysis, 153
Ricci-Curbastro, Gregorio, 152
Ricci flow, 151, 153
Riemann Hypothesis, 150
"Rise and Fall of the Quants, The"
    (Bernstein), 15–22
Ritchie, Dennis, 7
Rockefeller University, 6
"Rocket Science" (Bernstein), 121–132
Röhm, Ernst, 26
Rosen, Nathan, 131
Rosenbluth, Marshall, 39
Rosetta Stone, 95
Royal Air Force, 89
Royal Society, 75, 85, 91
Royal Society of Surgeons, 126
Rozental, Stefan, 29
Rubin, Robert, 16
Russell, Bertrand, 63
Russia, 52
    default on external debts, 21
    invasion of Afghanistan, 54
    production of first atomic bomb, 53
Russian Academy of Sciences, 150

S
Sachsenhausen-Oranienburg concentration
    camp, 27
Sakuntala (Kalidasa), 90
Salam, Abdus, 48
Salomon, 7, 17–18, 19
    Arbitrage Group, 19
Samuelson, Paul, 15
Sanskrit language, 84, 85, 86
    relationship to Greek language, 90–92, 103
    relationship to Proto-Indo-European
        language, 92–93
    William Jones' studies in, 84, 85, 86,
        91–93
Sarpi, Paolo, 67
Sceptical Chemist, The (Boyle), 67
Schliemann, Heinrich, 87, 88, 94
Schnabel, Arthur, 113
Scholes, Myron, 4, 22
Scholes, Myron, 4, 22. *See also* Black-Scholes
    equation
    autobiography of, 9
    at Massachusetts Institute of Technology,
        15
    as Nobel Prize winner, 7

"Schrödinger's cat," 157
Schwinger, Julian, 11
Science, depiction in novels, 122
Science and Hypothesis (Poincaré), 149
Science fiction, 122
Scientific culture, 122–123
Scioppius, Jasper, 59
Scripta Minoa, 101
"Self-fulfilling prophecy," 7
Shakespeare, William
    death of, 64
    Othello, 66
    The Tempest, 74
"Shakespeare of India," 90
Shawn, William, 48
Sheridan, Richard Brinsley, 80
Shimony, Abner, 131
Shipley, Anna Maria, 78, 84, 85, 86
Shipley, Jonathan, 74, 78
Shockley, William, 7
Shopping for Bombs (Corera), 52, 53, 55
Shore, John (Lord Teignmouth), 86
Siderius Nuncius (The Starry Messenger)
    (Galileo), 67
Smale, Steve, 151
Smith, Adam, 80
Smith, Alys, 63
Smith, Logan Pearsall, 61, 63–68
    autobiography of, 64
    The Life and Letters of Sir Henry Wotton,
        61, 64, 65–68
    Unforgotten Years, 63
    *The Youth of Parnassus and Other Stories
        of Oxford Life,* 64
Smith, Mary, 63
Smoluchowski, Marian, 12
Snow, C. P., 122, 145
Solvay, Ernest, 147
Sorbonne University, 12
Sorgenfrey, Robert, 163
Soros, George, 21
South Africa, nuclear weapons program,
    43–45
Space exploration, 41
Space quantization, 128
Spaceships, atomic bomb-propelled
    (Project Orion), 35–45
Space shuttle, 41
Spacey, Kevin, 165, 167
Spencer, George John
    education of, 75, 77, 78–79, 80–81
    father of, 78
    government career of, 77–78
    as Literary Club member, 80

Spencer, George John (*cont.*)
    marriage of, 78
    relationship with William Jones
        Jones as tutor to, 74, 75–79
        letters from Jones, 73–74, 76–77, 79,
            80–81, 83, 84–85
    as second Earl Spencer, 77, 84
Spencer, Georgiana, 82, 86
    as Duchess of Devonshire, 74, 79
    gambling debts of, 78
    marriage of, 78, 79
    William Jones as tutor to, 74, 76
Spencer, Henrietta, 78
"Spencers of Althorp and Sir William
        Jones: A Love Story" (Bernstein),
        73–110
Spin, of subatomic particles, 127–132
Sputniks, 37
SS *(Schutzstaffel),* 27, 28, 114, 116, 123
Stanford University, 129, 131
Steinbauer, Dieter, 110
Steklov Institute of Mathematics, 150, 154
Stern, Otto, 128
Stern-Gerlach magnets, 129, 130
Stock market, "beating the system" approach
    to, 159–168
Stock option pricing, 4, 10–11
    American call, 10
    Asian *versus* European, 17
    barrier, 12
    Black-Scholes equation for, 4, 7, 9–14, 19
    definition of, 9–10
    effect of arbitrage on, 11
    European call, 10
    fair, 159
    "in the money," 10
    "out of the money," 10
    "random walk" model of, 12–13
    "strike prices" of, 9–10
    synthetic, 11
    toy model of, 20
Stone, Harlan Fiske, 4
Stone, Marshall, 3–4
Stone, Vila, 4
Stoppard, Tom, 156
Storm Troopers, 26, 119
Strassmann, Fritz, 49
Subatomic particles, spin of, 127–132
Sub-prime loans, 18
Surgess, Jim, 165
Synopsis palmariorum matheseos
        (Jones), 77
Szpilman, Wladyslaw, 113–120
Szpiro, George G., 148

T
"Tales from South Africa" (Bernstein), 43–45
Taliban, 48
Tamil language, 93
Tao of Physics, The (Capra), 155
"Taxicab problem," 146
Taylor, Theodore ("Ted"), 37, 38, 41
Technical University, Delft, 51
Telescopes
    background noise in, 143
    of Galileo, 67, 68
Teller, Edward, 35–36
    "Plowshares" program and, 43
Tempest, The (Shakespeare), 74
Test ban treaties, 41
Théorie de la Spéculation (Bachelier), 12
Theory of general relativity and gravitation,
        22, 152–153
"Theory of Rational Option Pricing" (Merton),
        15
Thomas, Janine Ceccaldi, 133, 134
Thomas, Michel. *See* Houellebecq, Michel
Thomas, René, 133–134
Thompson, Ken, 7
Thorp, Edward O., 159–160, 163–167, 168
    Beat the Dealer, 166, 167
    "Fortune's Formula: A Winning Strategy
        in Blackjack," 165
3Com, 159–160
Three Degrees Above Zero (Bernstein), 6–7
Three-dimensions, 151–152
Thurlow, Edward, 83–84
Thurston, William, 152, 153
Topology, algebraic, 148
"Topology" (Bernstein), 147–154
Tory party, 82, 83
Toy, Bill, 16
Tradefin Engineering, 45
Transistor, discovery of, 7
Treasury bonds, 16
Treaty of Almelo, 52
Treblinka concentration camp, 115
TRIGA reactor, 36
Trinity College, Cambridge, 80–81
Tritium, 50–51
Troy, Greece, 94
Tsung Dao Lee ("T.D"), 5–6
21 (film), 165
21: Bringing Down the House (Mezrich), 167
Two-dimensions, 151, 152
Two-slits, 156–157
2001: A Space Odyssey (film), 41
Tycho Brahe, 59, 60
Typhus, 115

U
Ulam, Stanislaw, 36–37
Ultra Centrifuge Netherland (UCN), 51
Umschlagplatz, 115
Unforgotten Years (Smith), 63
"Unity of the Scientific Worldview"
        (Heisenberg), 31
University(ies)
    of Caen, 148
    of California at Berkeley, 131, 150
    of California at Irvine, 168
    of California at Los Angeles, 163
    of Cape Town, 5
    of Chicago, 3–4, 16, 19
    of Cincinnati, 94
    of Colorado at Boulder, 6
    Cornell, 124
    of Cracow, under Nazi occupation, 27–28,
        31–34
    of Islamabad, 3, 47, 48–49
    of Kiel, 26
    of Munich, 26
    of Orsay, 121
    of Paris, 149
    of Pennsylvania, 6
    of the Punjab, Government College, 48
    Rockefeller, 6
    Sorbonne, 12
    Technical University, Delft, 51
    of Turin, 107
    of Western Ontario, 131
UNIX operating system, 7
UPS (United Parcel Service), 42
Uranium
    enrichment of, 43–44, 50, 51, 52
    use in nuclear bomb production, 43–45
Uranium Enrichment Company (URENCO),
        51, 52
Uranium isotopes, separation of, 52
Urdu language, 84, 93

V
Vela satellite, 44
Venice, 58–59, 63, 64, 65, 66, 68
Ventris, Michael
    death of, 105
    deciphering of Linear B by, 88–90, 94, 95,
        97, 99–105
    "Introducing the Minoan Language," 89
Vetulonia, 108, 109
Vie Mode d'Emploi, La (Perec), 136
Vienna, 157
Von Braun, Werner, 45, 123

V-1 rocket ("buzz bomb"), 123
V-2 rocket, 122, 123–126
Von Weizsäcker, Carl Friedrich, 31
Von Weizsäcker, Ernst, 31

W
Wace, A. J. B., 95, 101–102
Walton, Isaak, 57
    Compleat Angler, 64
    friendship with Henry Wotton, 64
    The Life and Letters of Sir Henry Wotton, 61
    The Lives, 58
    Reliquiae Wottonianae, 62
Warsaw Ghetto, 113–120
    deportation of Jews from, 115–116
    escapes from, 116–118
    Jewish Council of, 116
    living conditions in, 114–115
    organization of, 114, 116
    underground army in, 117
    uprising in, 117
    Wladyslaw Szpilman's experiences
        in, 113–120
Wartenberg Castle, Cracow, Poland, 31, 32
Washington, George, 83
Wawel Castle, Cracow, Poland, 27, 31, 32
Weinberg, Steven, 6, 48
"What the #$"!?" (Bernstein), 155–157
What the #$"! Do We Know? (film), 155–157
Wheeler, John, 141
When Genius Failed (Lowenstein), 18, 20
Whig party, 81, 82
Whitman, Walt, 63
Whitney, Hassler, 149
"Whitney's trick," 149
Wilkins, Charles, 84, 90
Wilson, Robert, 7
Wotton, Henry, 57–69, 66
    as ambassador to Venice, 58–59, 63, 64,
        65, 66, 68
    art collection of, 68–69
    biographies of, 58, 61
    death of, 69
    The Elements of Architecture, 68
    friendship with Isaak Walton, 64
    knighthood of, 58, 59
    letters of, 63–68
    as Ottavio Baldi, 58–59, 65
    portraits of, 69
    as Provost of Eton, 57, 59, 62, 64, 67, 68
    relationship with Isaak Walton, 58
    relationship with Johannes Kepler, 60–61
Wurm, Leon, 120

Y

Yang, Chen-Ning, 5
*Youth of Parnassus and Other Stories of
    Oxford Life, The* (Smith), 64

Z

Zangger, Heinrich, 147
Zippe, Gernot, 52

Printed in the United States of America